IDEAS on the Nature of Science

Other books by DAVID CAYLEY

The Rivers North of the Future: The Testament of Ivan Illich

Northrop Frye in Conversation

The Expanding Prison: The Crisis in Crime and Punishment and the Search for Alternatives

George Grant in Conversation

Ivan Illich in Conversation

The Age of Ecology

IDEAS

on the Nature of Science

Edited by
DAVID CAYLEY

Copy-edited by Laurel Boone.
Cover image: H Berends, stock.xchng.com.
Cover design and interior page design by Julie Scriver.
Printed in Canada on 100% PCW paper.
10 9 8 7 6 5 4 3 2 1

Library and Archives Canada Cataloguing in Publication

Ideas on the nature of science / edited by David Cayley.

Interviews broadcast on the How to think about science
segment of the CBC radio show Ideas.
ISBN 978-0-86492-544-2

1. Science.   2. Science — Social aspects.   3. Intellectuals — Interviews.
4. Scientists — Interviews.   I. Cayley, David   II.Title: Ideas (Radio program)

Q171.I34 2009   500   C2009-904972-4

Goose Lane Editions acknowledges the financial support of the Canada Council
for the Arts, the Government of Canada through the Book Publishing Industry
Development Program (BPIDP), and the New Brunswick Department of
Wellness, Culture, and Sport for its publishing activities.

Goose Lane Editions
Suite 330, 500 Beaverbrook Court
Fredericton, New Brunswick
CANADA E3B 5X4
www.gooselane.com

Contents

IDEAS on the Nature of Science

Introduction

Ideas on the Nature of Science began life as a series of radio broadcasts called "How to Think about Science" that I presented on the CBC Radio program *Ideas* between November of 2007 and June of 2008. The series was inspired by an intuition that something had changed within that vast complex of ideas, institutions, practices, and products that we summarize and make manageable under the name of science, and changed quite dramatically. This was a change, I thought, not just in the practice of science, but in its public reception, in the way in which it is understood. "How to Think about Science" was my attempt to bring before my listeners some of the writers and thinkers who have had a hand in creating and promoting this new understanding of science.

Science has long been taken as the very definition of modernity. Historian of science Alexandre Koyré called the rise of modern science and the dissolution of the ancient cosmos "the most profound revolution achieved or suffered by the human mind." Herbert Butterfield claimed that the Scientific Revolution of the seventeenth century "outshines everything since the rise of Christianity and reduces the Renaissance and the Reformation to mere episodes. [It is] the real origin both of the modern world and of the modern mentality." American philosopher John Dewey saw science as synonymous with democracy. "The experimental method," he says, "is the only one compatible with the democratic way

9

of life." For him, science was "the organ of general social progress." Others have connected science with civility, skepticism, tolerance, and respect for evidence — all the virtues on which modern citizenship is thought to rest. Even George Grant, an anti-modern philosopher who considers "the modern paradigm of knowledge" to be a source of tragic alienation, still puts science "at the core of the fate of Western civilization."[1] In short, science has been thought of as the practice and the mentality that have made modern societies what they are and made them utterly unlike any previous form of society.

At the heart of these various pronouncements is a certain confidence that in speaking of "science" one knows what one is talking about — that the word refers to a more or less unified, more or less consistent body of ideas and practices, perhaps even to one overarching "scientific method." This confidence has been unsettled in recent years, as science has been exposed to a new kind of scrutiny. A characteristic note is sounded in the memorable opening sentence of historian of science Steven Shapin's book *The Scientific Revolution*: "There was no such thing as the Scientific Revolution, and this is a book about it."[2] It's a witticism that nicely captures a new attitude to the history of science. The phrase "the scientific revolution" — neat, pat, familiar — suggests, first of all, that this is a subject that we already know all about, and, second, that it is something that can be modelled as a compact and consistent event: a revolution. These are assumptions that Shapin wants to resist. The conventional label deprives the seventeenth century of too much of its strangeness and unfamiliarity. It suggests that we already know how the story is going to come out.

Historians of science, Lorraine Daston says in this book, have tended to be "people who . . . read the last page first." They told their story, in other words, as the unfolding of an end already implicit in its beginning. She and Steven Shapin belong to a generation of historians who have tried to rewrite the history of science with the controversy and contingency put back in. Anthropologists, sociologists, and philosophers of science have also tried to estrange themselves from familiar narratives. They have tried, anthropologist Allan Young says here, "to make explicit what is taken for granted." "Our impression was that we hadn't engaged closely enough with the work scientists . . . actually do," Simon Schaffer told me. "We relied almost entirely on what they said. We hadn't been to look." This "going to look" involved close observation in laboratories, hospitals, and field stations to find out how scientific knowledge is actually produced, warranted, and institutionalized. And it involved the study of controversies, present and past, where the facts were in dispute, the truth had not yet stabilized, and one could see scientific knowledge still, as it were, in its molten state. Finding out "what it is like to put the ship into the bottle" is the telling image used by Harry Collins, one of the pioneers of this type of study.[3]

These new studies have produced a new picture of scientific knowledge. There has been much more emphasis on the productiveness of science, on the ways in which science actively makes and remakes the world it lives in. Philosopher Ian Hacking, for example, points out that science doesn't just represent the world, it also changes it, creating phenomena that have never existed before. And much more emphasis has been placed on the diversity and heterogeneity of scientific practices. Philosophers in the first half of the twentieth century argued strenuously for the unity of science, which they saw as humanity's only effective bulwark against the obscurantism and irrationality of fascism. Today, the disunity of science is a more common topic. Thomas Kuhn set the tone with his idea that science is done within all-determining "paradigms" that lack common terms with the ways of thinking that come before and after them. Subsequent scholarship has tended to soften this austere view of knowledge as a sequence of locked rooms. Physicist/historian Peter Galison, for example, has argued that though the sciences do often speak mutually unintelligible languages, they communicate via "pidgins" that are elaborated in the "trading zones" that form at the boundaries between sciences or between theorists, experimentalists, and engineers within a science.[4] Even so, the picture which emerges is one in which it makes much more sense to speak of "the sciences" than simply of Science.

Along with this recognition of diversity has come a new appreciation of the limits to scientific knowledge. In these pages, for instance, you will find Dean Bavington's account of the collapse of Canada's cod fishery. Science once seemed capable of confidently modelling the ecology of the oceans, its synoptic gaze reducing codfish to a predictable population for which a maximum sustained yield could be reliably calculated. Today, a more humble science recognizes that the web of interactions within the oceans is too complex to allow this kind of prediction. Wendell Berry makes a similar point with regard to the way in which scientific agriculture overlooked the peculiar requirements of local ecosystems in its rush to expand production. And medical anthropologist Margaret Lock rejects the assumption of bio-medical science that there is a "universal body," and consequently a universal procedure for treating it, in favour of what she calls "local biologies."

Other thinkers I interviewed have studied the shadow that science throws on daily life when terms that make sense only within a network of precise scientific definitions leak out of science and into the vernacular. Geneticists know what they mean when they speak to one another about genes, but when genes enter popular parlance, they become what German scholars Barbara Duden and Silya Samerski call "pop genes." Pop genes have broken free of all the stipulations that make the scientific gene an intelligible object and become, in a sense, pure ideology, a bridge over the abyss that separates laboratory knowledge from everyday life. Believing ourselves to be the products of genes, Duden and Samerski say, transports us from the sensible world into a disembodied realm of risks. Ruth Hubbard underlines their point by pointing out that the genetic screening technologies that are now routinely used in pregnancy are "infinitely cruel to women" in the way in which they demand choice on the basis of impersonal and imponderable probabilities. Sajay Samuel, in a similar way, thinks that the prevalence of scientifically defined objects in politics disables political judgment. When politics revolves almost entirely around expert opinions about what is feasible, he says, there is little scope for citizens to express their judgments as to what is good or what is sufficient. His proposal for a return to common sense is echoed in a somewhat different key by David Abram. Abram's concern is with the habitual reduction of the sensible world to its supposedly more basic constituents. To this way of thinking, science gives us the real, our senses the merely phenomenal. He thinks this is entirely the wrong way round. The full dimensional world of our experience has the ultimate significance. Science, as an abstraction from this primary world, is useful and illuminating but ultimately secondary.

Restricting science to its proper sphere has also been the objective of those who have tried to demystify the political authority of science. British sociologist Brian Wynne, for example, has been interested in the ways in which political commitments structure ostensibly scientific judgments. Along with several other thinkers in this collection, he argues that science cannot currently answer the question of its own purposes. Science can create nuclear reactors or a race of transgenic marmosets whose skin glows in the dark; it cannot tell us whether this is a fitting thing to do. The answer lies outside of science. Yet these properly political questions, Wynne says, are still often disguised as scientific ones. He studied a public inquiry into the building of a new nuclear reactor in Britain and came to the conclusion that scientific rationality is very often used as "a ritual form of authority" in which the appearance of carefully investigating risks and assessing technical feasibilities is used to suppress questions of justice and propriety.

The reason scientific and political questions are still kept separate in this way, according to French thinker Bruno Latour, is due to the residual effect of what he calls "the modern constitution." At the beginning of the modern era, he says, after the devastation of the Wars of Religion, an attempt was made to establish a domain of reliable facts and to distinguish it from the realm of mere opinion. Science pertained to nature, the region of ascertainable fact, and politics to society, where opinion prevailed. Humans had agency and voice, nature was passive — science spoke for it. This was a fruitful fiction, Latour maintains, but it is long out of date. What was true all along has now become blindingly obvious — society and nature are inextricably mixed, and nature displays undeniable agency. Think, for example, of the international climate conference in Kyoto in 1997. Was this a political or a scientific gathering? Think of the response bacteria have made to antibiotics by evolving into what we call superbugs. Is this not a fateful intervention in the political world? "The 'Body Politik,'" Latour remarks in a recent essay, "is not only made of people," and the way scientists "make things public" is consequently no less political than the representation of people.[5]

Latour's argument suggests that it is not just the image of science, our way of talking and thinking about it, that has changed. The world itself has changed, and, in many ways, it is science that has changed it. In our interview, he makes the point, with characteristic wit, that if you had told René Descartes or Immanuel Kant that humans can influence the climate, they would have taken you for a believer in outmoded myths. But humanity *has* altered the composition of the atmosphere, and that has transformed not just the way we think about the relationship of science to society, but the relationship itself. German sociologist Ulrich Beck speaks of the emergence of a "risk society," by which he means, among other things, a society in which science can no longer predict or control its effects. The atomic bomb was not tested on people before it was dropped on them. There is no atmosphere other than our own in which we can measure the result of rapidly increasing the concentration of greenhouse gases. No one knows how genetic interference with plants, animals, and people will play out in the long term. Society has become, in effect, a scientific laboratory. For Beck, this radical novelty urges what he calls a second modernity, a reflexive or self-conscious modernity with institutions that enable us to begin to take responsibility for the consequences of this so-far uncontrolled experiment. Latour uses a very different terminology, arguing for a recognition that we have never, in fact, been modern. For him, modernity is a myth whose time is over. Its critical separations and analytical distinctions must give way to an ethic of care and composition which recognizes that all beings are now in the same soup and that "we can get rid of nothing and no one." The differences of language and approach are significant, but both thinkers point to the fact that science must be brought, as Latour says, "into democracy."

During the 1990s, the new anthropology, sociology, history, and philosophy of the sciences that I've been briefly summarizing engendered a reaction that became known as the science wars. The gist of the critique that was put forward in books like *The Higher Superstition* and *Fashionable Nonsense* was that "the academic left," as these books styled it, had betrayed the old left's devotion to science and Enlightenment and waded off into the swamps of relativism. The main bugbears for these writers were feminism, with its investigation of the ways in which masculine bias has coloured the practice of science, and the many varieties of the view that scientific knowledge is "socially constructed," that is, produced in social settings under social assumptions and not just innocently discovered. A lot of the public controversy seemed to turn on caricatures and taunts — "if you're so sure the laws of gravity are socially constructed, maybe you'd like to jump out of my window" — but, when I look back on it, the whole affair looks to me less like a debate and more like a last-ditch effort to save the credit of an obsolete image of science.[6] This is not to deny that science had lost some of its innocence and some of its aura — its high-modern mystique — but it was hardly because a handful of radicals in science studies had pulled back the curtain and revealed the great Oz as a mere mortal. What was at stake was nothing less than what Bruno Latour calls the "modern constitution," with its strict distinction between nature and society and its elevated view of science as an authority exempt from any touch or taint of politics. No wonder deep anxieties were stirred.

Today, at the end of the first decade of the new millennium, the situation feels very different. It seems easier to discuss science quizzically without being immediately asked to declare whether you are a friend or an enemy or whether you "believe" in Boyle's Law or Maxwell's equations. Yes, there are still people who think the earth is six thousand years old, as there are still scientific zealots who accord science the place of true religion, but, in general, I would say that a lot of what was threatening in the new studies of science of the 1980s and 1990s is now much less so. The time seems right, then, to look at the new accounts of science that emerged during the last half of the twentieth century and at the marked differences amongst them. *Ideas on the Nature of Science* is the record of one such attempt. It reflects the vicissitudes of preparing a radio series — not all the people I would have liked to interview were available — and it reflects the partialities of my reading and my acquaintance; suggestions from listeners as to all the people I had left out would have easily supplied a second series of comparable length and comparable interest. The interviews were recorded between the fall of 2006 and the spring of 2007. Some episodes of the full broadcast series have been omitted to conserve space.[7] But, even so, I think the collection provides an interesting and representative sample of a field whose riches no one reader could ever exhaust. The points of view represented are diverse and sometimes contradictory. Themes recur, but no attempt has been made to reduce them to a common denominator. My hope is simply that readers will find here resources with which to think about science. Much depends on it.

NOTES
1   Alexandre Koyré, *Metaphysics and Measurement* (London: Taylor and Francis, 1992), p. 20; Herbert Butterfield, *The Origins of Modern Science 1300-1800* (London: G. Bell and Sons, 1949), p. viii; John Dewey, *Underlying Philosophy of Education*, in *Later Works*, Vol. 8, *Collected Works of John Dewey*, ed. Jo Ann Boydson (Carbondale IL: Southern Illinois University Press, 1969-91), p. 102, and *Democracy and Education*, in *Middle Works*, Vol. 9, *Collected Works of John Dewey*, ed. Jo Ann Boydson (Carbondale IL: Southern Illinois University Press, 1969-91), p. 239; and George Grant, *Technology and Justice* (Toronto: House of Anansi, 1986), p. 9.
2   Steven Shapin, *The Scientific Revolution* (Chicago: University of Chicago Press, 1996), p. 1.
3   Harry M. Collins, *Changing Order: Replication and Induction in Scientific Production* (Chicago: University of Chicago Press, 1992), p. 145.
4   · Peter Galison, *Image and Logic: A Material Culture of Microphysics* (Chicago: University of Chicago Press, 1997).
5   "From Realpolitik to Dingpolitik," in *Making Things Public: Atmospheres of Democracy*, ed. Bruno Latour and Peter Weibel (Cambridge MA: MIT Press, 2005). See also *We Have Never Been Modern* (Cambridge MA: Harvard University Press, 1993) and *The Politics of Nature: How to Bring the Sciences into Democracy* (Cambridge MA: Harvard University Press, 2004).
6   See Paul R. Gross and Norman Leavitt, *The Higher Superstition: The Academic Left and Its Quarrels with Science* (Baltimore: Johns Hopkins University Press, 1994 and 1998) and Alan Sokal and Jean Bricmont, *Fashionable Nonsense: Post-Modern Intellectuals Abuse of Science* (New York: Picador, 1998). The invitation to test the social construction of the laws of gravity comes from Alan Sokal's unpublished letter to the *New York Times*, which he posted on line; see *http://www.jwalsh.net/projects/sokal/articles/skl2fish.html*. You can get a bit of the flavour of the rest of the debate from a published exchange between Bruno Latour and Ashraf Noor; see *Common Knowledge*, 8, no. 1 (2002), pp. 71-79 or *http://muse.jhu.edu/journals/common_knowledge/v008/8.1latour.html*.
7   A full transcript of "How to Think about Science" is available by writing to *ideas@cbc.ca* or calling the CBC Shop at 1 800 955-7711.

Knowledge Is
an Institution

SIMON SCHAFFER

I think there are two standard images of what the sciences are. One image is that scientists are absolutely special people, that they're much more moral and much more virtuous and much, much cleverer, and that they do things that are nothing like what anybody else does. And on the other hand, there's an equally powerful public image of science, which is that science is organized common sense, that it's just cookery raised to a fairly sophisticated art. Those are the two dominant images of public science in our culture, and neither of them is right.
— Simon Schaffer

In 1985, a book appeared that seemed to sum up a new approach to the history of science — one that had been gradually taking shape since the 1960s. There had been precursors, of course, but up to that time the history of science, broadly speaking, had meant biographies of scientists and studies of the social contexts in which scientific discoveries had been made. Scientific ideas were discussed, but the procedures and axioms of science were not put directly into question. *Leviathan and the Air Pump* involved a more searching interrogation of the history of science. Subtitled *Hobbes, Boyle and the Experimental Life*, the book's avowed purpose was "to break down the aura of self-evidence surrounding the experimental way of producing knowledge." This was a work, in other words, that wanted to treat something obvious and taken for granted — that matters of fact are ascertained by experiment — as if it were not at all obvious, that wanted to ask, how is it actually done, and how do people come to agree that it has truly been done?

The authors of this path-breaking book were two young historians, Steven Shapin and Simon Schaffer, and both have gone on to distinguished careers in the field they helped to define, science studies. I spoke with Simon Schaffer at his office at the Whipple Museum of the History of Science at Cambridge, where he teaches. We talked first about why science came under a new kind of scrutiny during the period when he was just beginning his studies, the 1970s. This was a time, he said, when many modern certainties were shaken, and there was no greater modern certainty than the authority of science. In this atmosphere, a new generation of scholars began to ask new questions. No longer content just to take scientists at their word, they wanted to see for themselves how science is made.

SIMON SCHAFFER

Our impression was that we hadn't engaged closely enough with the really lived work scientists and technologists actually do. We'd relied almost entirely on what they said. We hadn't been to look. So significant groups of social scientists, mainly in Britain, interestingly, began to work alongside scientists in labs, in field stations, in research clinics, in zoos and botanic gardens — to follow the scientists around to try and look at what they did. We were using field methods, in other words, borrowed entirely from the field sciences, except this time, instead of looking at lemurs, we were looking at physicists. Instead of looking at Trobriand Islanders, we were looking at Californians.

DAVID CAYLEY

These observations yielded a picture of science that varied dramatically from the image that philosophers of science had put forward. Science, according to these philosophers, was essentially reason in action. Schaffer and his colleagues came to a very different conclusion.

SS: It seemed that we were dealing with groups of very skilled crafts-people. People who were ingenious and clever and skilful, who had lots of what we call tacit knowledge, who were very well trained, who were more like engineers than they were like priests. They didn't have very much bigger brains. Their skulls looked very similar to the skulls of other people. They didn't seem to be doing anything terribly special in terms of methods. They didn't seem to be much more skeptical than the rest of us. They didn't seem to be constantly making bold conjectures and then desperately trying to falsify them. They seemed to be the ingenious manipulators and managers and artisans of carefully designed workspaces, whether in labs or in the field. And we wanted, I think, to say that that was how it was.

DC: Scientists, in the eyes of Schaffer and his colleagues, began to look less like oracles and more like skilled carpenters. Their knowledge was not the very voice of nature but a human product, something that had to be made and then maintained. This insight turned the received view of the relationship between science and society upside down. Formerly, science had been seen as social only when it was wrong: social interests distorted and corrupted knowledge, but true knowledge was immaculate, untouched by human hands. Now, the sciences began to be understood as inherently social.

SS: Knowledge is an institution, and it should be analyzed as such. What groups of people say about the world, how they come to agree on it, and how they find out how things are — these things all have an institutional quality and should be analyzed the way other institutions are analyzed. That meant, for example, that it was extremely unpromising, to put it mildly, to suppose that social principles are only acting when folks get things wrong. So, for example, it didn't look remotely plausible to say that Isaac Newton thought that there was an inverse square law of gravity acting instantly at a distance through empty space between the centres of distant bodies because that was the fact of the matter: there *was* an inverse square law acting instantly from the centre of one body to another through empty space. Whereas Leibniz disagreed because he was German.

That is, it didn't seem appropriate to us to explain the truth, what we think is so, in one way, and to explain what we don't think is so in a completely different way, as though there are these things called social forces which wreck our ability to see how things are. What we learned was that there are social institutions at work to produce what we know and indeed to produce what anybody claims to know at any particular period. It seemed just very odd that only if you escape from society can you see how things really are. That didn't seem right to us. It seemed to us, and it still seems to me, that people in social groups build their knowledge like they build other institutions.

It follows that you should analyze how they do it in the same way that you analyze the other institutions people build. That meant that we were trying to think as hard as we possibly could about how people come together to build the institutions by which they live. Some of the most important institutions by which they live are the things they hold to be true about the world. That meant, therefore, that it would be a really, really good idea to look at scientific controversies both in the present and in the past. Look at controversies in the present, in which you don't yet know what the right answer is. Then you'd be looking at rival groups trying to make their knowledge claims into institutions that everyone could join. If you didn't yet know the right answer, if you could follow a controversy, or what was called science in action, you could see how people come to believe what they believe and know what they know.

To use a phrase from, I think, one of the most important sociologists of scientific knowledge, Harry Collins, you could see how the ship gets in the bottle. Once it's in the bottle, you can't imagine how it got there. You can't see the little hinges in the mast. You can't imagine the tweezers and the glue. It just looks as though it's always been there. What we wanted to do was to see how ships get put into bottles. As they're being put into bottles, they look messy and complicated and contingent and possibly chaotic, and maybe they won't fit. Maybe it'll all fall to bits and some different vessel will end up in some different container.

We wanted to do this, as I said, for the past as well as the present. So, while my colleagues in Bath and Edinburgh and Paris and elsewhere were studying current fights in neuroscience and biochemistry and astrophysics and nuclear technology, we also wanted to look at comparable episodes in the past, where the job, we thought, might be even harder. The reason the job might be even harder is that we know the end of the story. We know what's going to happen in the end. But imagine that you don't. Imagine that you go back three hundred years, and you don't know who's going to win. You don't know what the right answer is yet. Can you follow through a fight in the past as if you didn't know who had right on their side? And then try to unfold the process by which what's the case about the world was established and institutionalized? That's what we set out to do.

DC: Simon Schaffer has undertaken a number of studies in which historical controversies are used as a window through which one can observe science in the making, the ship as it's being put into the bottle. The most famous of these is the book *Leviathan and the Air Pump*. Co-written with Steven Shapin, it concerned Robert Boyle, the most noted English experimentalist of the seventeenth century, and his contemporary, Thomas Hobbes, who disputed Boyle's claim that experiment is the most certain road to knowledge. Here, in outline, is the story.

SS: Once upon a time, there was a very wealthy Irishman called Robert Boyle, who was born into great wealth and great social prestige in the 1620s; he is known as the son of the Earl of Cork and the father of modern chemistry. By the 1650s-1660s, a time of political crisis and cultural revolution in Britain, he'd established himself first in the city of Oxford and then, after the restoration of King Charles II, in London, in a very fashionable part of town, as one of the masters of a new kind of project — experimental philosophy is what he called it. It involved using elegant and complicated machinery to find out how nature worked by performing experiments. These experiments mobilized a whole new series of techniques which now seem, or might seem, to us to be absolutely obvious. But they were new in that period, and so they had to be explained and defended.

First of all, Boyle designed and commissioned complicated engines and machines. The most complicated, and the most prestigious, was an air pump. The reason he was interested in the air was because one of his heroes, the great royal physician, William Harvey, had demonstrated that the blood circulates in humans and animals and that it passes through the interior surfaces of the lungs. Something, William Harvey said, enters the blood from the outside world through our lungs; the question is, what? Why does the blood circulate? That was the chief research question of Harvey's followers and therefore of Robert Boyle.

Boyle's genius, you might say — and again, this has become absolutely self-evident for scientists now — lay in his realization that if you want to know about the properties of something, it's a really good idea to get rid of it and then see what difference that makes. Since they couldn't do experiments on the moon, where there's no air, they built an air pump instead and put animals inside this machine to see what happened to them if they didn't have air. That was the research agenda.

Now, to go very quickly, the reason that's a new kind of approach is that Boyle was mixing up two categories which, up till then, had been radically opposed to each other: the category of nature and the category of art, or engineering as we might say. Roughly, before the 1600s, natural philosophy, the knowledge of nature, was understood to be knowledge of how nature normally behaves, of how it commonly is.

So, for example, we know that bodies made of earth fall in straight lines towards the centre of the earth, because that's what they normally do. We know that bodies composed of fire move in straight lines away from the centre of the earth, because that's what they normally do. Men like Boyle were arguing that, no, the way to find out how things are is to create singular, strange, mechanically produced instances. Stop observing nature as it normally is. Start producing effects which you can isolate and analyze. That's a huge shift in how to find out about nature.

But in order to do that, other techniques were needed as well, and in *Leviathan and the Air Pump*, we focus on two of those. One was that Boyle recruited witnesses, people he brought together in the room where these experiments were being done, people who would agree that what Boyle said was going on *was* going on and who would later back up his printed account of the experiment. This activity of witnessing became very, very important for the new science, because it meant that knowledge was collective. A group of people stood committed to a particular claim. One starts to see how, in order for experiments to work, they have to become socialized. They have to be communal. They have to be a kind of group activity. That was the argument.

Then, finally, how do you describe experiments when you write down the account and then print it and distribute it? We were struck by the way in which Robert Boyle, in particular, wrote; when you read his stories, it's as if you're watching what he's seeing. We call that virtual witnessing. This new kind of writing brought you into the presence of what was being described and allowed you to imagine you were a witness, too. And, since witnessing an experiment was so important, but everybody couldn't be brought into the room, this literary technique, as we called it, was a way of multiplying the number of people who could be imagined as taking part in making the fact.

So we had working with a machine, bringing people together to see the machine working, and then writing down what was seen so that readers were, at least in imagination, also present. If all that was in place, you produced what Boyle called a matter of fact, something that was so authoritative and so plausible that no one could deny that it had happened. Finally, there was an invitation to others to repeat what Boyle had done. You were invited, though it was expensive and difficult and complex, to build your own pump and check what he'd done.

DC: Why, at that time and in that place, did it become so important to produce a matter of fact?

SS: Well, there's a question with which I think citizens in the early twenty-first century will be very familiar. We live in a time of war and global struggle, when it's extremely hard to know what's really happening in the world. Rival powers divide people's allegiances, there's a huge clash of ideologies — between states, between religions, between different groups of people — thousands of people are being killed for their beliefs, and one doesn't know where to look for secure, well-grounded authority. In these circumstances, any group of people that comes along and says, look, we have a method, we have an approach, we have a program which will give you undeniable truths, on the basis of which you can then reason without fighting, and you can come to agreement through negotiation — that group of people is offering a solution to a major social problem. It's still a social problem for us.

We still live — in fact, we intensely live — in a world dominated by a crisis of trust. We aren't sure which experts to believe. Perhaps it's because there are too many experts. Perhaps it's because there aren't enough. That was the situation in the middle of the 1600s in Britain, and indeed in most of Western Europe. This was an epoch of crisis, of war, of religious and political discord. Robert Boyle and his friends made that point explicit. They argued: we live in an age in which men use the sword rather than reason to come to agreement, in which there's violent dispute; we need a way, we need a whole enterprise which will produce undeniable truths and then allow people to see how to argue without killing each other. The experimental philosophy explicitly offered itself as a recipe, not just for producing truth, but for displaying how to argue without violence or catastrophe. Thus the solution to the problem of knowledge, which Robert Boyle and his friends offered, their method of producing agreement, became a solution to the problem of social order. How do you get citizens to behave? How do you get a social group to come together to argue, negotiate, and agree? Making the experimental philosophy exemplary was a political, as well as a scientific, achievement. And they said so.

DC: Trust — establishing his credibility, as we would say — was a very important matter for Robert Boyle. Shapin and Schaffer's study of how Boyle gained assent is one of the most original aspects of their book. Very often it is skepticism, or a disbelieving attitude, that is treated as the hallmark of science. The motto of the Royal Society, of which Robert Boyle was a founding member, was (and is) *Nullius in verba*, on no man's word. John Locke, Boyle's contemporary and associate, says, for example, that "in the sciences everyone has only so much as he *really* knows. What he believes only, and takes upon trust, are but shreds." Schaffer argues, on the other hand, that the critical achievement of science lies not in its skepticism, but in its ability to create and maintain trust.

SS: The basic question which collective public knowledge always has to solve, in every culture, is, who should I believe and why? Almost everything everyone knows about the world they know on trust. Almost all our knowledge is testimony. Very little of what we believe we know about the world is based entirely and absolutely on our own experience, and the social order requires that kind of mutual trust. One of the models that we've inherited in our world is empiricism. Empiricism is the philosophical position that my knowledge comes from what I experience. That's a problem, because most of what I know and believe I haven't directly experienced. I've experienced it through others. I've experienced it not immediately, but mediately. Adam Smith, the great Scottish economist of the eighteenth century, puts this wonderfully in one of his lectures. He says, if you reckon up what you know about the world, you'll see that most of what you know has been purchased, like your shoes and socks, from those whose business it is to put these goods on the market. I think that's still absolutely true. If I reflect on what I know about the world, what I think is true, almost all of it is based on what I've heard or been told or read, not what I've directly experienced.

I think a useful slogan is, the Western natural sciences work very well, partly because they organize trust extremely efficiently, not because they organize skepticism and doubt extremely efficiently. We tend to think that scientists are the people who don't believe anything, who question everything, who don't rely on anybody else's belief, who believe only what they see with their eyes and feel with their fingers. And there's certainly something in that. But, by and large, when one visits or works in a laboratory or an observatory or a botanic garden or a hospital or a zoo, one quickly realizes that, no, actually the really impressive aspect of the natural sciences in our culture is that they seem to be very good at knowing who to believe, who else to believe, who else to trust. They have these extraordinarily long-range networks which distribute trust and authority relatively efficiently.

If you look around a working laboratory, likely it'll be full of machinery and equipment that has not been made by the people working in the lab. They have to rely on the people who've made those devices, or made those observations, or calibrated the contents of those bottles, or designed that piece of software. So a laboratory is a place connected through immensely complicated networks to the labours of many, many other people. If those people's work can be trusted, then what goes on in that lab can be trusted. The network of science and technology is a very good system, so it seems, for getting it right about testimony and trust.

DC: Only a handful of people actually witnessed Robert Boyle's experiments with the air pump. Even fewer ever tried to repeat them. Some who did got different results. This was one of the reasons why Thomas Hobbes was not impressed by Boyle's claim that experimental science was the royal road to civility and social peace. Hobbes thought the proper model of knowledge was geometry, with its carefully defined axioms and demonstrable proofs. The air pump, he said, was "of the nature of a pop gun," something which children use, though, he allowed, "[more] costly and more ingenious." Such a toy could not provide demonstrable knowledge. Its operation was too unreliable, its operators too potentially self-interested. It was an argument history had forgotten but which Steven Shapin and Simon Schaffer took very seriously.

SS: We were excited and fascinated by the dispute which broke out in 1660-1661 between Hobbes and Boyle, because it seemed to raise a lot of the questions we were also interested in. Hobbes had a very deep objection to the kind of claims the experimenters were making. His principle objection was that they were exaggerating the security of their operation. They said, he thought, that if you get a group of people together and you show them a singular phenomenon, some experiment or other, then they will come to agree, and they will never dispute about the result they've seen. Hobbes thought that was incredible. That's to say, he didn't believe it. He did not believe that if you get a group of people together, and what they're looking at really matters, they will effortlessly agree about what's going on in the world. His experience, as he understood it, was different. In 1660, he was seventy-two years old, and he could think back over a life, in Britain and elsewhere, which showed him that if folks' interests clash, they will clash. You need much more powerful methods for bringing people to agree and behave and obey than simply showing them a machine or a phenomenon and then hoping that, because they've all seen the same thing, they'll all say the same thing. Hobbes simply didn't believe the ambitious claims the experimenters were making. He couldn't see why, if these are issues of major concern to people, they would suddenly all give up their interests and their passions and all say, yup, you're right, we're wrong, that's how it is, we all agree, and peace has broken out. Hobbes was, you might say, more pessimistic, you might say more realistic about human nature. He didn't foresee, could not foresee, a world in which people would come to an easy agreement just because they'd been shown something. That was one major argument he made.

A second kind of argument that Hobbes put forward was this: just because something happens in one place, how do we know it happens elsewhere? You claim, on the basis of what happens in the back of your house in Pall Mall in London, that there can be empty space in the world, or you claim that, because you've seen a stone fall in a chamber, therefore all stones fall like that everywhere. How do we know that? Show me. Show me it happening everywhere. Show me it happening always. You can't do that. You can't build a knowledge of the universe from what you see on one occasion.

DC: So there's no such thing as experimental philosophy, in Hobbes's view, and there can't be such a thing. It's not philosophy.

SS: Yes. It isn't philosophy. Unless you've already reckoned — and it's Hobbes's favourite word, reckoned — unless you've already reckoned on what you know and what must be the case about the world, simply seeing what happens on one occasion at one time in one place is not going to be enough to convince you. It's not how you build up universal knowledge. What you have to do is to analyze and reason. In that sense, Hobbes is the rationalist in this fight. Hobbes admired the power of human reason. He admired the way in which we were able to reckon up what we know, what must be the case about the world, rather than building up what he would have seen as a house of cards from these special privileged experimentally produced instances.

Those two criticisms fit together. His first criticism is that experiment cannot compel people to agree, and people won't agree unless they're forced. His second criticism is, how do I know, simply by looking at an experiment, that it will always be thus? If you can't show me that it will always be thus, I don't have to agree with you. And if I don't have to agree with you, probably I won't agree with you. It is politically dangerous, Hobbes finally argues, to offer a recipe to produce peace if you know the recipe won't work. These were powerful criticisms and they needed answers, and they elicited from Boyle and his allies answers which are very interesting because they had to spell out more and more of what they were assuming about the world and of the basis of the claims they were making. They had to spell out their views of social order. They had to spell out their views of religion. They had to spell out how they thought these techniques of agreement and trust were supposed to work.

Hobbes, who was possibly the most irascible human being of the seventeenth century, went on fighting for the rest of his life, more or less. What's interesting in the end is that that fight was then, to a large extent, forgotten, forgotten not only by historians of the sciences, but also, more interestingly, ignored by people who were fascinated by Hobbes. Because it was embarrassing, perhaps, that this great rationalist, the chief theoretician of modernity (which Hobbes, in many ways, is — secular, rational, progressive, modern, a brilliant analyst of politics and of philosophy), was on this occasion, so it seemed, on the wrong side. Why? Because he was fighting the experimenters. He was fighting men like Robert Boyle, and that's not nice.

One of the things we set out to do was to go back to a fight which was, in a way — I don't want to exaggerate this, but was, in a way, hidden from history and bring it back to light to take what was at stake seriously. If you take the critics of experiment remotely seriously, that's a high-risk strategy, because it might look as though you're on their side, whereas what you're trying to do is to bring up for analysis the assumptions that people make when they make claims about how we should find out what there is in the world.

DC: *Leviathan and the Air Pump* was a controversial book. It was widely read outside the new field of science studies and is probably still the most frequently cited text that field has produced, but it also attracted the charge that it was anti-science. Schaffer thinks this was the result of the point of view he and Shapin adopted. They wrote, in a sense, as anthropologists, distancing themselves from taken-for-granted assumptions about science, trying to get a fresh look at things usually taken as too obvious to even question.

SS: I think we were accused of being subversive and anti-scientific because, no doubt, the kinds of stories we wanted to tell and the kinds of questions we wanted to ask seemed unfamiliar or troublesome for certain kinds of images of what the sciences are. I think there were two standard images which were in trouble as a result of what we seemed to be saying. One image, which was very important in the past of the sciences, is that scientists are absolutely special people, that they're absolutely different from the rest of us, that they do things nothing like what anybody else does, that they're much more moral and much more virtuous and much, much cleverer, and altogether a different group of persons. That was a very important theme in the long history of public science. Lots of people have said that, but we were saying, no, actually, from that point of view, the people in the past and the people in the present whom we're trying to understand don't seem to be morally or ethically or cerebrally different from the rest of us.

On the other hand, there's an equally powerful public image of science, an image which is subscribed to by many great scientists of the past and the present, which is that science is organized common sense, that there's nothing special going on in the sciences. It's what everybody does all the time. It's just cookery raised to a fairly sophisticated art. That's equally been a way of defending the value of science. Those two images — scientists are absolutely different from us, and all they're doing is what we do — are the two dominant images of public science in our culture, and neither of them is right. They both have enormous problems associated with them. If you make the scientists into some kind of clerisy, some kind of priesthood, this has extremely bad effects on politics and knowledge and culture. It's also, it seems to me, an entirely inaccurate way of understanding what scientists do. If, on the other hand, you say that they're not doing anything different, they're not doing anything special, there isn't a special family of techniques, nothing terribly different goes on in laboratories and clinics, you just don't understand what does go on there and what is special about the way these complicated institutions are organized.

So, between those two extremes, we wanted a third way. The friends of the two extremes attacked us with, I have to say, rudeness, violence, inaccuracy, misquotation, and the threat of unemployment in several cases. It was a very unpleasant period indeed. Several of my friends either lost their jobs or were threatened with losing their jobs on the grounds that they were the enemies of the sciences. It was a very unhappy period.

DC: This period, roughly from the late 1980s to the late 1990s, became known as the science wars. Historians like Simon Schaffer were accused of constructivism, that is, of portraying scientific knowledge as no more than a vulnerable human construction. But why, he wonders, should scientific discoveries be thought less true or less noble just because they are seen to have been made by human hands?

SS: By showing exactly how the institutions of our society come into being and how they work and why they're so effective, we were accused of making them ineffective . . .

DC: . . . by taking the mystery out of them?

SS: Perhaps. Certainly my friend David Bloor would say that what was going on in that stupid debate recapitulated a debate which happened at the end of the eighteenth and the beginning of the nineteenth centuries between people who were literalist about the Bible and people who wanted to write its history. People who pointed out that the Bible had been written by human beings and that the texts of scripture should be analyzed in the same way as other texts were accused by their enemies of atheism, of denying the truths of scripture, of subverting the institutions of the Christian religion. There are some very interesting similarities between that fight, mainly in the German lands around 1800, and these fights, mainly taking place in New Jersey and Wisconsin in the 1980s and 1990s. I think that's a very interesting analogy. Demystification is often treated as catastrophic undermining.

DC: The science wars was a name that lumped together a variety of happenings: a few academic black-listings, a handful of polemical books, a lot of caricature. Like the guests of the legendary Procrustes, who boasted of an unusual bed into which anyone would fit but achieved the feat by violently lengthening or shortening the guest, people seeking a revised understanding of science were chopped or stretched to fit the narrow bed of anti-science. Simon Schaffer views it in retrospect as a panic in which the purging of certain outworn myths was taken as something far more threatening. Today the controversy seems to have died, or at least been superseded. One reason might be broad changes in the sciences themselves, changes that can hardly be attributed to the work of a few controversial scholars. In fact, the very idea of what a science is has changed dramatically.

SS: When I was a student, the model science was theoretical physics. That was the pattern science. All other sciences aspired to the status of theoretical physics. When you learned the philosophy of the sciences, the examples were taken from theoretical physics, from Newton, from Maxwell, from Einstein. And that meant that when you were writing about the sciences, you had in your mind the idea of what an abstract mathematical physicist does as the pattern you were trying to explain. Advancing complicated mathematical hypotheses, making predictions which are then tested by experiments carried out in controlled conditions in carefully designed laboratories: that was the image of the sciences that we had, and all other sciences were imperfect versions of that. That was what we were told.

I think that over the last thirty to forty years our image of what we're trying to explain and what the sciences are like has completely changed. One way of understanding how it has changed is that for us now the pattern science is something like agronomy or field botany. In other words, we've switched from thinking of the pattern science as theoretical physics to thinking of the pattern science as a field science. Field sciences involve an immense amount of travel and the construction of field stations in the wild. They involve moving people around from place to place. Ecology and agronomy, I think, are now the pattern sciences. The plant trials that people do, to see if genetically modified organisms work in one place or in another — are they safe here or there? — and the field trials people do on new forms of transportation, those are more like what the sciences are.

Now, if that's true, you'll start asking completely different questions. One change is that instead of asking who came first, which is the traditional question for the historian of science — was it really Isaac Newton who was the first to describe universal gravitation? Was it really Albert Einstein who was the first to formulate the principle of special relativity? — instead of those questions of time order, you'll ask questions about space order: I see this technique works here; how does it work there? I see that these plants grow here; how did they ever come to grow there? How do drugs get tried and moved around; how do clinical trials work? Those are the questions that historians and sociologists of science, I think, are much more possessed by, in a way. Not only do they, as it were, switch from chronology to geography, from who came first to how does it happen to work over there, but they also, therefore, are much more interested in why people agree than why they disagree.

DC: New questions have produced a new image of science as an extensive and effective social institution. This more realistic, less mythological portrait of the sciences needs to be conveyed to the public, Schaffer argues. Then, when Science with a capital S pronounces, citizens will have some idea of how such pronouncements actually take shape.

SS: A good way of thinking about it is to think a little harder about the phrase "the public understanding of science." I don't know about the Canadian situation, unfortunately, but in Britain this is a common expression. It started to come into use in the mid-1980s when it was perceived that insufficiently large numbers of people were entering the sciences as a career, which I think, sadly, is absolutely the case. At that time, too, opinion pollsters could easily find vast numbers of people on the streets of London who thought that atoms were smaller than electrons or that the Earth goes around the moon or something like that. And that seemed disturbing and worrying, which it is.

So a movement for the public understanding of science started. But the phrase "the public understanding of science" is intrinsically ambiguous. It might mean, how many people understand the view of the world that science gives us? It might mean, how many people understand how scientists produce a view of the world? There's a slippage between those two senses of public understanding of science.

I still remain convinced that without one you can't have the other. You've got to make sure that people have a very good understanding of how scientists achieve an understanding of the world, and through that, you can get people to understand the view of the world that the sciences produce — its worth, its status, and its contents. Without that, you don't have citizens. You don't have intelligent citizens who can take part fully in the debates that really matter, whether they're about nuclear disarmament or global warming. You similarly don't have a clear model of how to distribute trust in our society. I think that's one of the most important issues in our world today concerning the sciences. It's not that we don't trust anybody, it's that we don't know who to trust. We're surrounded by experts. We have experts who tell us how to eat and how to drink. We have experts who tell us how to get married and when. We have experts who tell us how to invest and how not to invest. The world both in North America and in Europe is stocked full of expertise. This is not a period when there's been a crisis of expertise or a loss of deference. On the contrary, you could argue that people defer to unprecedentedly large numbers of experts. I didn't even know until a few years ago that I was supposed to ask advice about how to organize a wedding or how to educate my children, although now, of course, one does.

If we had better, more accurate, more effective models of how expertise and trust work in our culture, and above all of how trust and expertise are distributed with the sciences and with technology, I think this would be a crucially progressive move in our politics as well.

A Story Without
an Ending

*What is really remarkable about science is its ability to create novelties.
And these novelties are not just at the level of new discoveries and
new theories; they involve whole new ways not only of understanding
knowledge, but of getting knowledge and of being a knower.*
— Lorraine Daston

Through most of modern Western history, natural science has been
idealized. Science has meant enlightenment. Think of Alexander Pope's
lines on Isaac Newton: "Nature and nature's laws lay hid in night./ God
said, let Newton be, and all was light." Science has meant progress:
Pandit Nehru, India's first prime minister, told his country's scientists
in 1961 that "science alone could solve India's problems." And science
has served as a model of democratic community. "It gives us room to
differ without animosity," said the first historian of the Royal Society,
Thomas Spratt, and allows "contrary imaginations without any danger
of civil war." Enlightenment, progress, civility — these were the gifts of
science and what made the West the very template of modernity.

During the last thirty to forty years, this received wisdom has been
very substantially revised and the idealized portrait of science replaced

by a more shaded and nuanced image. One of the places where the new history of science is being written is the Max Planck Institute for the History of Science in Berlin. Housed in an elegant and airy new building in a leafy suburb of the city, it is home to approximately a hundred scholars whose research extends from medieval cosmology to the role of experiment in nineteenth-century German gardening to the ways in which medical technology has reshaped the contemporary boundary between life and death. The director is American Lorraine Daston. I interviewed her in her office at the Institute, and she told me that there was a time when she would not even have dreamed of a hundred historians of science under one roof. When she was a graduate student at Harvard in the 1970s, she said, the history of science was more a collection of strays from other disciplines than a discipline in itself. But a crucial challenge had already been issued. In 1962, philosopher-historian Thomas Kuhn had published *The Structure of Scientific Revolutions*, the book that put the previously unusual word "paradigm" on everybody's lips. Kuhn rejected the assumption of a continuous linear progress in science. In doing so, he framed the question with which Lorraine Daston's generation grew up: how to write the history of science as something other than a triumphant progress to a foregone conclusion.

LORRAINE DASTON

Imagine you're the kind of person who cheats when you read mystery novels, and you read the last page first to find out who did it. Then, when you read the rest of the mystery story, you know that everything is building toward this climax. You read it in a very different way than the person who has to retrace all of the red herrings that have been placed in the reader's way by the author to throw that reader off the scent of the real villain.

We, the historians of science, were the people who always read the last page first. We knew how the story ended. Moreover, we were complicit in this with the scientists themselves, who wished to have a story about why we believe what we believe now and why it is right. What the scientists knew, and the historians did not know, at least not in quite so visceral a way, was that everything — or most things — which we now believe in science will be overturned, if not in a generation, then in two generations. This is the pathos of science. It is the progress of science, but it is the pathos as well. The kind of history of science I'm talking about provided reassurance. It said, it is inevitable that we should believe what we believe now; there are good reasons; indeed, there are no other possible positions one could have. It was that deep reassurance which the scientists needed even more than the historians.

But Kuhn convinced people with very powerful arguments that this was a betrayal of history and also a betrayal of science. It was a betrayal of history because you can't write history knowing the end; you'll lose what it felt like to be in the seventeenth century, the eighteenth century, and so on. You lose the past. You make the past over in the image of the present. And it's a betrayal of science, because it negates what is most distinctive about science, which is its enormous creativity, its enormous capacity for renovation. So that message had been digested. Then a second, even deeper message of historicism came with Foucault: you know those objects that you think have histories? Well, there are a lot more objects out there that have histories. Like sexuality. Something that was thought to be a constant for all humanity for all time for all cultures turns out to have a fascinating, full of surprises, full of hairpin turns history. Whoda thunk it?

DAVID CAYLEY

Michel Foucault expanded the very idea of history. Historians had always known, as the novelist L.P. Hartley famously said, that the past is a foreign country, a place where they do things differently. But Foucault went further. He suggested that history is interrupted by what he called epistemic breaks, deep ruptures in the very meaning of the basic categories by which we make sense of things. Thomas Kuhn said something similar with his idea that knowledge is organized in distinct paradigms. Both thinkers made the past a much stranger place than it had seemed when it was treated as mere prologue to a predestined present. Lorraine Daston built on these foundations.

One of the projects with which she began was a history of wonder. Working with her friend and colleague, Katherine Park, she investigated the ways in which people thought about monsters, marvels, and prodigies between the years 1150 and 1750. Published as *Wonders and the Order of Nature*, the book argues that around 1750, what the authors call "a profound mutation" took place in the way scientists and intellectuals generally thought about themselves. At that date, they say, wonder became "vulgar," "a disreputable passion" beneath the dignity of a sober scientist, "redolent of the popular, the amateurish, and the childish." But before that time, Daston says, a lot can be learned about the history of science by considering how people viewed things that lie beyond the boundary of rational explanation. Here's her précis of the book.

LD: It's about things which undeniably happen. We're not talking about miracles. We're talking about marvels — things which undeniably happen but which do not fit neatly into any classification scheme you have at your disposal and which leave you, therefore, in one of two states that turn out to be peculiarly closely related to one another. Either you're in a state of wonder, which is a positive state — there's something pleasurable about the surprise. I mean, if you were to see the aurora borealis, that's probably the state that you would be in — or it leaves you in a state of horror because you feel there's something portentous in what has happened. Literally portentous. It's an ill augury. Nature is out of joint. Katherine Park's and my departure point was really a question about what kind of phenomena trigger these very specific emotional responses and how is it that one tips over into another? What are the historical and intellectual preconditions that make wonder tip over to horror, or curiosity tip over into wonder? And that's what the book is about.

DC: Would you give an example of such tipping?

LD: One of many examples I could give is a sermon that was preached in the early part of the seventeenth century in Plymouth, England, in which a minister berates his congregation for their wonder at Siamese twins who have recently been born in the parish. He says, this should be an object of fear and horror to you. God is trying to send us a message. Repent! Repent before it is too late, instead of gawking as if it's Mayfair

Day. So this is a case where the congregation has tipped from horror into wonder, and there's a certain amount of consternation on the part of the religious powers that be that this has happened. Or take the example of wonder and curiosity. For us, it seems absolutely axiomatic that if you see something like the aurora borealis, after that moment of gaping astonishment, your first reflex would be, what causes such a thing? Why doesn't it happen all the time? That question is in no way intuitive for the people who study nature in the Middle Ages. First of all, they think curiosity is a sin, and secondly, they think that wonder should be reserved for religious experiences, not for natural experiences. They see wonder as very much akin to what we would call awe and fear. Therefore, their first reflex is not in the least curiosity. That would be almost blasphemous. Rather, a certain reverence would follow that experience.

DC: Curiosity is a sin?

LD: Yes.

DC: Following Augustine?

LD: Following Augustine, and even before that, there's the Greek idea of *periergeia*. The first meaning of curiosity is to butt your nose in where it doesn't belong. It means trying to find out whether your neighbour's wife is having an affair. It means, still worse, trying to pry into the secrets of the prince, and, by extension, the secrets of nature's prince, God. Thus to investigate the things that God has hidden from us — this is an act of impudence at the very least. Busybodyness is perhaps the mildest form of it, and blasphemy is the worst form of it.

So, no, curiosity has a really bad rep until about the sixteenth century. Then you begin to see a change. There is a wonderful passage in Francis Bacon's *Great Instauration* where Bacon says, look, it's true that Adam and Eve sinned, but they sinned by seeking moral knowledge, not by seeking natural knowledge. The reason why we have to pry into nature in order to find out how things work is that God is playing a game of hide and seek with us. Bacon cites a passage from the Bible, from the Old Testament, about King Solomon, which says that it is for God to make secrets and it is for a king to find them out. So: a kind of challenge, a friendly challenge that God had put before Solomon. Then you begin to get a whole set of other instances like that, which attempt to at least neutralize curiosity, and eventually — it's one of the few cases in the history of the vices and virtues in which a previous vice becomes an unequivocal virtue. But you can see why it was a vice.

DC: Curiosity was never one of the deadly sins, was it?

LD: No, it wasn't.

DC: Or did it assimilate into one of the deadly sins?

LD: Well, that's a very good question, because it is related to *superbia*, the Latin word for pride, and it's considered to be part of the sin of Lucifer before he fell. His pride is his wish to revolt against God, but it's also his desire to know things that he has no business knowing. So, yes, curiosity implies a kind of vaulting ambition that's allied to pride. And the other thing about it is that it's related to *avaritia*, it's related to avarice. It's one of those things which isn't self-limiting. There are sins like gluttony, where you're eventually going to reach a point of satiety. You're just going to eat yourself sick and you won't be able to have another ice cream sundae. And even with lust there's a point of satiety. It just can't go on. But curiosity is never sated. It's like a thirst which is never slaked. That's why Hobbes says that it's more pleasurable than any carnal pleasure: it is inexhaustible. However, passions that are inexhaustible are extremely menacing to the social order. We call it addiction in our language of the vices and virtues. It makes people completely incalculable. People who are in the grip of a gambling passion cannot stop themselves. In Romantic literature, the novels written about scientists are all about people who can't stop. They know they should. They know they're going to kill off their nearest and dearest if they don't stop. They can't stop. There's a great novel by Balzac, called *La recherche de l'absolu*, about a man named Balthazar Claës, who is a chemist. Not an alchemist. He is a student of Lavoisier.

He orders all his instruments from the best instrument makers in Paris. He is a pillar of the community. But he's addicted to chemistry, and he destroys his family. He destroys himself because he can't resist just one more chemical experiment. That's why curiosity is a vice. It's like *pleonexia* in Greek, which was this kind of passion that doesn't stop. It knows no limits.

DC: Curiosity was a medieval vice that became a modern virtue, even if writers like Balzac and Mary Shelley continued to ponder its shadow side. This re-evaluation of curiosity took place during the Renaissance, and Daston thinks one of the main reasons was Europe's dramatic opening to the world at this time.

LD: I think it has to do with a profound recognition, during the period in which it happens, the sixteenth and seventeenth centuries, of novelty. You have to imagine a society which is just overwhelmed by a tsunami of novelty, the voyages of exploration, the novelties of the Far East and the Far West. There are new stars in the heavens. There are new religions. There are new commodities. The European market is being saturated with things that people have never seen before. They turn up in *Wunderkammer*, for example.

DC: *Wunderkammer* are . . .

LD: . . . chambers or cabinets of curiosity. The German term literally means cabinets of wonder, chambers of wonder, and these are basically a way of making three-dimensional and visible all of the neat new stuff.

DC: What might be in one?

LD: Ostrich eggs, paper money from China, a canoe from Lapland, a two-headed snake, stuffed, a cherry stone carved with the Lord's Prayer. You name it, it's in there — a piece of marble from Tuscany which looks like a landscape, a fly in amber. There's this overwhelming experience of a world suddenly crowded with new things. And curiosity of course is the passion for assimilation. It's the appetite for the new, and it's that appetite which is kindled and stoked by, for example, the printing press. I mean, the very word "news" is a creation of this period because it's possible through broadsides, one-page sheets that are disseminated everywhere to keep people abreast of what's happening now. A whole new way of reading is created, which is to read a whole lot of things quickly rather than reading just one book over and over again. I think that's what triggered this re-evaluation of curiosity in that period.

DC: This re-evaluation of curiosity under the impress of novelty led to a new attitude towards wonders. Daston finds an instance of this new attitude in the writings of Francis Bacon, the Lord Chancellor of England in the early seventeenth century and the exponent of a new science.

LD: Bacon's position on curiosity is tightly interwoven with an awareness of wonders in the world. Bacon tries to clear the boards. He says, look, we've got to start all over again. The best and the brightest have been wrong for centuries. We've made a terrible mistake with regard to how nature works. We have to start all over again. But how are we going to start all over again? he asks. We're going to have to start at the foundations, which is not explanations of how things work but finding out just what is there in nature in the first place. Let's take inventory.

He says that this natural history, this reformed natural history, is going to have three parts. It's going to have the part that it's always had, namely what usually happens, nature in its normal course. Then we want to include another part, which is nature wrought, nature modified by human beings, as when we make machines or tables and chairs or when we weave fabric.

Finally, he says, we want a third part, and this is going to be a history of nature out of course, a history of monsters. I want a collection of wonders, of everything that happens which is new, rare, and unusual. The reason I want that is because it will unsettle our self-evidences, the things we take for granted. The great problem with the study of nature heretofore has been that people have leaped from a few commonplace examples to the loftiest of generalizations, and then they've gone fearfully astray. We've got to pay attention to the exceptions in nature as well as to the rule if we're going to have any chance of having laws which will encompass everything in nature. Moreover, he says, the wonders of nature afford the nearest passage to the wonders of art. If you study nature's marvels, you'll get hints about how we can make our own marvels.

The history of technology, Bacon says, is a history in which things actually get better. He says, look at the compass. Look at the printing press. Look at gunpowder. Those are three things the ancients never had. That shows that we moderns are at least in this respect superior. How come our study of nature, how come our science isn't superior? We've been flooded with novelties of art. Where are the novelties of nature?

DC: Bacon's question received a resounding answer during the course of the seventeenth century, as Park and Daston are able to show in *Wonders and the Order of Nature*. Increasingly, wonders shifted from the periphery to the very centre of scientific inquiry.

LD: The first scientific societies are founded in the middle decades of the seventeenth century. There's the Academia dei Lincei, in Rome, of which Galileo is a member. There's the Academia Naturae Curiorsorum, those who are curious about nature, which is founded here in Germany. There's the Royal Society of London, and there's the Académie des Sciences in Paris. If you look at the annals of these first scientific societies, they are chock full of reports about wonders: three suns seen in the sky; two-headed baby born in Sussex; rain of blood in Bavaria; army of ants marching in formation wearing what looked like baseball caps in Strasbourg. They read like the *World Weekly News*. They read like the *National Enquirer*.

DC: "World War II bomber found on moon" sort of thing?

LD: Exactly. Or three duck hunters shoot down angel by mistake. When Katie and I were writing this book, we used to buy the *National Enquirer* and write it off on our taxes as a research expense. I mean, we were absolutely convinced that the guys who wrote for the *National Enquirer* were simply looking at back issues of the *Phil. Trans.* [*Philosophical Transactions of the Royal Society*] from 1670 and, you know, rewriting them in modern colloquial English. That's what the first scientific societies are concerned with. Not exclusively. There are also what we would consider more properly scientific papers, but they are cheek-by-jowl with news of the latest two-headed cat, reported by none other than Robert Boyle, by the way. Or Leibniz is sending in a report to the Paris Academy of Science about a dog who can bark out the words *chocolat* and *café*. It's not just the hoi polloi who are submitting these reports, it's the luminaries of the scientific revolution. And this is part of the Baconian program. These people are saying, look, Bacon told us to compile a natural history of pretergeneration, and that's what we're doing.

So you get, as I say, the heyday of wonders. Then, around the 1730s, a reaction sets in. It happens first in France and much later in Germany and in Britain. The French really slam on the brakes and say that amidst all of this blooming, buzzing confusion, all of this variability, we really must look for the regularities of nature. They begin to develop an ideology as well as a practice of systematically screening out reports of wonders to the point where, in the late eighteenth century, when reports of what we would call a meteor shower reached Paris, the Académie des Sciences refuses to credit it. Then, in 1802, there is a meteor shower within two kilometres of Paris, at which point even Laplace has to admit that there is a such a thing.

But it smacks too much of the wondrous. I think this is part of a really profound philosophical response to the problem of belief. In the seventeenth century, you can make a mistake, but the mistake usually involves incredulity rather than credulity. People who are writing as natural philosophers will say, only a country bumpkin would refuse to credit the reports we've had from the New World about some new, strange kind of fish. People who have travelled, people who have read, realize that there are more things in heaven and earth than are dreamed of in your philosophy. The sophisticated, intellectual attitude is a kind of omnivorous openness toward the most wondrous things, the things that are being reported in the *Philosophical Transactions of the Royal Society* or the *Acta Eruditorum* or the very well named *Miscellanea Curiosa* in Germany. By the 1730s the pendulum is swinging, and the new threshold for belief is set very high. The worst sin you can commit as an intellectual is credulity, is gullibility, is believing too much. And that's where it has stayed fixed pretty much ever since.

DC: The sober, skeptical, almost ascetically disbelieving scientist remains a familiar image today. More surprising, therefore, is the taste for the wondrous that Lorraine Daston and Katherine Park discovered among the first generations of modern science: Leibniz and the talking dog, Boyle and the two-headed cat. It unsettles a familiar image. Earlier historians of the era of Boyle and Leibniz have applied labels that connect this time to the present: the Scientific Revolution, the early modern period. Both names make us feel we know where we are. The Scientific Revolution is obviously pregnant with modern science, the early modern can only be on its way to the modern. Park and Daston emphasize the radical break between the omnivorous curiosity of the first scientists and the cool skepticism of their successors. In doing so, they give the past back some of its surprise.

Wonders and the Order of Nature is just one example of how the history of science is currently being rewritten. There are fewer grand narratives in which the present is clearly inscribed in the past. There is more emphasis on novelty and discontinuity, and yet Daston recognizes that there still have to be stories that tie things together. Her latest book is an attempt to tell such a story. Co-written with Peter Galison and titled simply *Objectivity*, it's a history of objectivity which examines how this idea has changed from the eighteenth century to the present.

LD: What we wanted was not a bird's-eye point of view, but something like what you have when you're tracking a storm. There are some phenomena, like a storm front, which, if you're a localized observer here in Berlin on the eleventh of June, 2007, you can't possibly reconstruct. You have to have a vantage point that allows you comparisons with what's happening in Oslo or what's happening in the Canary Islands or what's happening in Madagascar in order to reconstruct this phenomenon. Or, to alter the metaphor, take the distribution of a species — an isolated observer can't tell you the distribution of blackbirds in

Western Europe. We have to have far-flung observations. That's the level at which the book is written in an attempt to create a narrative. It's a narrative which has a very different shape than the familiar narratives of the history of science, in that it's neither progressive nor episodic in the Kuhnian sense. It's neither punctuated by revolutions nor evolutionary. It is a narrative in which novelty erupts but doesn't displace the old. It modifies the old, but it doesn't displace it. We'll see whether or not that grips readers as a narrative. The proof of the pudding will be in the reading.

DC: Thomas Kuhn argued that scientific knowledge is ordered by what he called paradigms. When Nicolaus Copernicus demonstrated that the earth revolves around the sun, he inaugurated a new paradigm. And paradigms, according to Kuhn, are generally incommensurable. That is, they possess no common measure. So, for example, Newton and Einstein may both use the term mass, but its meaning is coloured very differently by their respective systems. This picture of revolutionary breaks between different world pictures is what Daston means by the Kuhnian sense of narrative as episodic. She and her co-author were not satisfied with Kuhn's story of abrupt discontinuity, but neither did they want what she calls an evolutionary narrative in which ideas are continuously refined and improved. What they came up with instead was a view in which styles of thought do not succeed each other but rather overlap. They argue that different conceptions of objectivity respond to different fears as to how thinking can be misled. The story begins in the eighteenth century, when the word "objectivity" was not yet used in its current sense and the great fear was a diseased imagination.

LD: The malady that the eighteenth century fears most is twofold. Thinkers are worried that you're going to be overwhelmed by the pell-mell sensations that are streaming in upon you. Unable to sort them out, you'll either be lost in the chaos of them or you'll shut down entirely. You'll blinker yourself. You'll abandon experience and retreat to some airy castle of the imagination, and there you'll succumb to some *esprit de système* and build a fantasy world of your own because you've shut out experience. That's what they're most worried about, and they take precautions. One of the precautions they take is to say that you've got to have a very strong faculty of reason. What reason does to experience is to shape it, to prune it, to select it, to find the hidden order underneath all of that surface variability. What reason does to the imagination is to tell her to stay in her place. Imagination is always described, even in languages that don't have grammatical gender, as feminine. She's always pictured as a coquette. The job of reason is to resist the blandishments of the seductress, the imagination. So that's the danger, and those are the precautions you take within this framework. The greatest error you can make is to mistake the appearance of things, all that surface variability — this plant with an odd leaf, this skeleton with a broken rib — for the real truth of nature, which is some underlying prototype.

DC: The ideal of truth to nature seeks the typical rather than the transient form of things. It fears being misled by variation, so it seeks an underlying archetype. But in the nineteenth century, a new ideal begins to appear. Daston and Galison call it "mechanical objectivity," and they connect it with a new division of the world into subjects and objects.

LD: If you're in a world which is divided up into objectivity and subjectivity, what you most worry about is not that you will be overwhelmed by experience, but that you will overwhelm experience. You are worried that your subjective self will leap up and project its own theories and hypotheses and expectations onto nature and deform it. Hence the need for all those forms of self-restraint, all those mechanical rules that you must follow. You may decide that, instead of making a drawing of the crystals that you're studying under the microscope, you'll take a photograph. Or if you're an astronomer, you'll decide to use a mathematical formula to reduce your data rather than your judgment. These are all precautions against this overactive self.

DC: Objectivity in the nineteenth century became almost an ascetic practice. The noted German physiologist Rudolf Virchow told his scientific society in 1877 of his prolonged effort to, as he said, "de-subjectivize" himself. The self threatened scientific knowledge. Virchow spoke of his opinions, his theories, and his speculations as things to be shaken off in the pursuit of what he called "an objective mode of being." This was a moral as much as a philosophical stance, according to Daston and Galison. In fact, they argue that any intellectual stance is always simultaneously an ethical one. Styles of knowing are moralized, and scientists moralized them very differently in the nineteenth century than they had in the eighteenth.

LD: When you look at the kind of moralization that takes place amongst these eighteenth-century naturalists, it's very much in the Neo-Stoic or — perhaps better — Aristotelian mode of what ethicists call a virtue ethic. It's an ethics which is based upon habit. It's not based upon the exercise of the will. It's based upon forming regimens of the right sort through long experience in the way in which you train your memory, you train your attention, you train judgment and reason. If you look at the kind of ethics which is associated with objectivity, it's very Kantian. It posits a self, a dynamic self, which is organized above all upon the exercise of the will. And it is the will which is paramount in the moralization of objectivity, scientific objectivity.

The will plays very little role in these eighteenth-century injunctions to elevate reason above the imagination. The will is considered to be inefficacious in subduing the imagination. The only hope you have is to fortify reason in the same way that you would fortify a muscle: by exercising it strenuously each and every day.

DC: The practice of reason in the eighteenth century and the dominion of the will in the nineteenth century had to make room in the twentieth century for a third style of knowing. Galison and Daston call this third mode "trained judgment." It came to the fore when scientists began to recognize the role of intuition in science.

LD: They say it would be foolish not to avail ourselves of our considerable human resources in pattern recognition. Wittgenstein made this idea famous with his idea of family resemblances, but it's everywhere in the early twentieth century — in some reputable places and in some not-so-reputable places — for example, in racial atlases. But it's everywhere, and the idea is that, just as you are an expert at recognizing the physiognomy of human beings, so you can be trained to be an expert at detecting patterns which no algorithm, no mechanical rule could ever capture. That's what we should be cultivating, at least in certain sciences and certain disciplines.

In our typology, each thought style matches a particular faculty. Truth to nature corresponds with the faculty of reason, objectivity with the will, and trained judgment with perception. Truth to nature and mechanical objectivity have not disappeared from the field, but they've been joined by this new approach which is that we don't care what your will does. In fact, we're not even interested in your consciousness. We think this exercise of judgment probably goes on mostly in your unconscious. So instead of straining every nerve, every muscle in a heroic act of will to get things right, we think the best thing you can do is go to sleep. Maybe then answers will come to you. This has concrete ramifications — in scientific pedagogy, first and foremost, but also (and this of course is the red thread through the book) in the way in which scientific images are made.

DC: Collections of images in scientific atlases provide much of the evidence on which Daston and Galison base their history of objectivity. They point at the end of their work to a new kind of image emerging from the still largely unfamiliar realm of what is called nanotechnology. Nanotechnology means manufacturing at a molecular scale, and it produces images in which the boundary between science and engineering, knowing and making has begun to dissolve.

LD: Something new is happening. With truth to nature or mechanical objectivity or trained judgment, what was always at stake was representation. The kinds of images that were produced under the influence of each one of these epistemic virtues may have looked different, but they were always meant in some way to be faithful to nature. What seems to be happening now, as we speak, is that the very idea of representation is being thrown overboard. Visualizing something, making an image of it, has become an integral part of making the object itself. It no longer makes sense to talk about a pre-existing nature which we then try more or less faithfully to represent. It only makes sense to talk about the presentation, the creation of an object, its manufacture through the very act of visualizing it.

This also has ramifications in the way in which scientific disciplines are organized. More and more, they're organized along lines that used to be more typical of engineering firms and departments, where the concern was to make objects and to sell objects and to present objects. It also means that the boundaries between nature and art, not just science and art, but nature and art, have been almost completely eroded. This is happening not only in nanotechnology, but also, for example, when one thinks about genetics now. It no longer makes sense to talk about nature and nurture in an era which is talking about designer babies and about the manipulation of the genome. We are in a position to dissolve the boundary between nature and nurture.

Something analogous seems to be happening at the boundary between the scientific image and the artistic image. There's no longer any pejorative association attached to certain kinds of manipulation of images to make them appear beautiful, as any journal cover of these days would show. Now, this has not happened without protests. There are many people, especially editors of scientific journals, who are extremely worried about what can be done to images with Photoshop or to statistical data with statistical packages, computer statistical packages, and they are trying to set limits to that — to create rules. These are the voices of mechanical objectivity protesting against this new development.

But the way in which this story ends — and you had asked about narrative before — the way in which this story ends is to suggest that there is continuing creativity, that what is really remarkable about science is its ability to create novelties. And these novelties are not just at the level of new discoveries and new theories; they involve whole new ways not only of understanding knowledge, but of getting knowledge and of being a knower. The reason why we offer this glimpse into what is happening now, and it can, of course, be only the most provisional and tentative of glimpses, is to drive home the message that we have not told a story with an end.

DC: Daston and Galison's story without an end is also a story in which the past continues to be present. The various forms of objectivity may emerge in sequence, but they don't just disappear when new forms emerge. Botanists continue to prize truth to nature, astronomers to employ trained judgment in the assessment of anomalous data. This idea of a plurality of overlapping styles of knowing challenges the popular image of science as a singular form of knowledge. People sometimes speak, for example, as if the whole genius of science consisted in the application of an unchanging technique called the scientific method. If one takes this view — that science is a simple, consistent object, everywhere and always the same — then the very idea that objectivity has a history can be somewhat threatening. And Lorraine Daston says that she does sometimes encounter critics who think that writing a history of something is tantamount to debunking it.

LD: If you tell people you are writing a history of objectivity, you often get a bristling response: don't you believe in it? My reflexive answer is, of course I believe in it. Would I write the history of something I don't think exists? But I think that that response represents a widespread opinion: to write the history of something like objectivity is, ipso facto, to try and make it disappear, to try and claim that it's an illusion, an artifact, an ideological construct of some kind. I am very puzzled by this reflexive identification of historicism with relativism. I don't understand why the fact that something has a history would in any way imply anything about its truth or falsehood. I'm interested in a history of the very deep categories of rationality, and I think two things which are perhaps surprising from the standpoint of most philosophy of science. First of all, I think that rationality is a plural. There are many different forms of it. Second, I think that it has a history, and it's an extremely productive history. That is, science is creative in the way that we were just talking about; it is spewing forth, like a fountain, all of these new forms. Perhaps not every ten minutes, but at least every century, it seems to come up with something of really earthquake-like novelty. That is the project on which I am embarked. But I think that this is a history which is not cumulative, in the old sense of a building, where you put one brick upon another and the tower gets higher and higher. It is cumulative in the sense in which one accumulates stuff. It's more like your attic, you know. Oh, there are my notes from sophomore year in college, and here is that waterbed. Remember the waterbed, the era of the waterbed? This history is more like accumulating stuff. Then the real mystery — and this is something which, I think, neither I nor others have yet confronted — is the mystery of, okay, so you have all of these separate forms of knowing, ingenious, deep forms of knowing — the experiment, objectivity, mathematical deduction; the history of science is littered with them — but how do they come together? How are they braided together? One of the most interesting things about the way science functions is its ability to integrate the novelties it spews up so creatively. The novelties cannot be allowed to perpetually collide with one another. That would bring the enterprise to a grinding halt. They have to somehow be integrated. And that history of how they come to be integrated has not yet begun.

For a decade or so, historians have been working on histories of evidence, histories of demonstration, histories of observation. This is going on now, here at this institute, but we do not have a history of how they all fit together.

DC: Coherence is the challenge that now faces those who have succeeded all too well in the ambition with which they set out: to overturn simplistic and self-serving narratives in the history of science. Daston's book with Katherine Park on wonders begins with a quotation from Michel Foucault that sums up this ambition. Foucault suggests, in the passage they quote, that a more curious gaze on the past can break down present certainties and produce new, less constrained possibilities — what he calls "a multiplication of paths." But today Daston wonders about the limits to multiplication.

LD: The idea was that if you could undo the inevitability, the apparent inevitability, of the way we think now, then we would actually have more choice. To give an example, there's a wonderful essay by Anne Fausto-Sterling, who is a biologist at Brown University, called, I think, "The Five Sexes." She points out, on the basis of anatomical, endocrinological, and sociological research, that the idea that there are only two sexes is

really a crude oversimplification. There are at least five of them. So —
imagine! — who would have thought that? Isn't that an emancipatory
possibility?

I think that this type of approach was naive. The idea that one had
at first in reading Foucault was that history would be spread before you
like an enormous smorgasbord, in which you would pick and choose.
You would say, ah, ancient Greece, we're going to take their tranquil and
sovereign attitude toward homosexuality, at least male homosexuality,
but we're going to reject their attitudes towards slavery. And we like
this little bit of the Middle Ages here, but we'll reject the rest. I think
that smorgasbord approach to composing a society will not work. What
makes all of these parts cohere into a liveable society? I don't think you
can create chimeras and expect them to have traction and reality. I don't
think you can take very disparate ways of knowing and just assemble
them side by side and assume that this will work as a viable way of being
rational. So there's another challenge, which is, why do some patterns
cohere and not others? That strikes me as the road ahead.

Local Biologies

MARGARET LOCK

Cultural anthropologists have tended to assume that there is a universal body. My position is that there is significant biological variation among different human populations.

— Margaret Lock

In 1993, medical anthropologist Margaret Lock published *Encounters with Aging: Mythologies of Menopause in Japan and North America.* The book explores dramatic differences in the way women experience menopause in each place. Such variation is usually taken as purely cultural, but, in her book, Lock makes a surprising suggestion: she proposes that there are biological differences between Japanese and North American women. Culture, in her view, doesn't just interpret biology, it also shapes it.

Margaret Lock is a professor in the Department of Social Studies of Medicine at McGill University, and I interviewed her at her home in Montreal. Before we got to her reflections on what she calls "local biologies," I talked to her about another path-breaking book of hers, published in 2002, called *Twice Dead: Organ Transplants and the Reinvention of Death.* The book explores a region that's new in our time

— the intersection of nature and culture. Modern societies, until relatively recently, tried to keep these categories distinct. It's obvious that all human artifacts blend nature and culture in some way — that's been true since our ancestors painted the walls of their caves — but the two could still be distinguished. This boundary no longer holds. Human beings now produce their own climate, freely alter the genetic constitution of plants and animals, saturate the air with invisible voices. Which is nature, which culture? The proportions in which they are mixed can no longer be sorted out. We're surrounded by what French philosopher Bruno Latour calls "hybrids" in which nature and culture, biology and politics have fused — the ozone hole, the Oncomouse, the test-tube baby. Lock is interested in how people live in this hybridized, redesigned nature. In *Twice Dead*, she studies the different ways in which Japanese and Western cultures have responded to a new kind of death — what we now familiarly call brain death — and to the removal of organs from apparently living bodies that have been declared brain dead. This new reality was relatively easily accepted in the West, but in Japan, where she has conducted much of her research over the years, it created scandal and rejection. Our discussion began with the piece of medical technology that lay behind the redefinition of death: the ventilator.

MARGARET LOCK
The ventilator came into use in the early part of the twentieth century, but it became widely used with polio epidemics because people who had contracted polio often needed assistance with breathing. The early ventilators were great big lumbering machines, very different from the kind of things one that one sees these days in hospitals.

DAVID CAYLEY
One has heard of an iron lung. Is that an early model?

ML: That's an early ventilator, yes. They started to be widely used. They were much more widely produced and were used in emergency medicine centres, now called intensive care units, with anybody who, for one reason or another, was having difficulty breathing. What then happened was that it became clear that certain people were dying, or at least were entering a state of irreversible loss of consciousness, while on these ventilators, and there was concern about when you could call them dead enough to remove the ventilator. When can you pull the plug on these people who are betwixt and between? They were taking up a lot of space in intensive care units, and people were feeling at a loss as to how to handle this. So it was eventually agreed that there would be this condition, initially known as irreversible loss of consciousness, which would be recognized as a condition after the diagnosis of which you could then remove the ventilator.

DC: The advances in ventilator technology that created this new kind of living dead coincided during the 1960s with the first successful internal organ transplants. In late 1967, there was a furor of publicity when a flamboyant South African doctor by the name of Christian Barnard performed the first heart transplant. Drugs that could suppress the body's reaction to foreign tissue became available soon afterwards. The stage was set for the realization of a long-standing medical ambition. But there was a problem. Living organs can only be obtained from living bodies. The solution was to redefine the irreversibly comatose people who were being kept alive on ventilators as dead. Brain dead. And it was this new category that the Japanese found hard to accept.

ML: You have to agree that this hybrid person who is no longer a person, who is betwixt and between, is both living and dead. You have to be able to agree, as it was agreed in North America and almost all of Europe, that he or she can be counted as dead, that the person no longer exists, and therefore that you can commodify this body. You can take, with permission of course, you can take organs from it.

In Japan, for many years, they were unable to come to this conclusion, and even now, for many Japanese, this is not a comfortable conclusion. Many people will agree that the condition is no doubt irreversible, but for them, somebody who is lying there on a ventilator, breathing and warm — the flesh is warm to touch, the skin is pink, the person looks alive — for very many people this cannot be counted as the end of human life. You should wait. You should wait longer until there's every evidence that there will be no recovery, evidence that satisfies family members entirely. This very often happens — not always, but very often — four or five or six days after the actual neurological diagnosis of brain death. But by that time the organs are no good for procurement — possibly a kidney, but certainly nothing else.

This debate went on in Japan from the 1960s through to 1997, and it became their biggest ethical problem — way bigger than anything like abortion, which has never been a huge ethical problem there. The Japanese Legal Association was opposed to the recognition of brain death. They were absolutely certain there would be all sorts of court cases and upsets about recognition of brain death. They also tend to vie with the medical profession on many matters. Everybody recognizes that the Japanese family, even with modern medicine and big hospitals, tends to be the ultimate decider as to when death has happened, rather than the doctor. The doctor will intimate and explain and then allow the family to come to a decision among themselves, this indeed is where we stand, and to make decisions about what they're going to do. They are much less pressured by intensive care practitioners in Japan than they would be here, and they are also only very tentatively asked about organ donation.

There's also, of course, a long tradition in Japan of recognizing life to be diffused throughout the whole body — it's not a country where Cartesian ideas took hold very strongly. The centre of the body doesn't just reside, as many of us might think these days, in the brain. That makes it very counterintuitive for many people to think of this as the end of the person, that there is no person still existing after the diagnosis of brain death.

DC: Brain death and organ harvest are instances of the astonishing novelty that techno-science creates. The human condition is put under permanent renovation. People had known, everywhere and always, how to recognize and respond to death. But now death had lost its definite boundaries and would have to be certified and adjudicated by experts. The body's frontier became less certain as well, and in the process, entirely new spiritual and social realties were created.

ML: Getting a new organ is a transformative experience. You know, in the case of a cadaver donor, that the generosity of some unknown other has saved your life. Some people accept that in a matter-of-fact way, and one could become very psychological and say they repress all the social aspects of this and just get on. I've got my organ and I'm going to get back to work, sort of thing. Maybe some people are like that, but, when you interview people both in North America and in Japan, you find that for a very large number it's not like that at all. They continue to wonder about the donor. They continue to wonder who it was and what kind of personality he or she had. Some people feel that they've taken on new personalities. Some people find that they like foods that they used not to like. A good number of people actually talk to their new organ to try to make it feel comfortable in their body. Give it little chats about, you know, we're going to be all right today, and this is what we're going to do today, and so on.

So a huge amount of animism is connected with this enterprise, although the medical world would not normally want to talk about that and damps it down. But even some doctors feel this way. In the book, I reported a case of one heart surgeon whom I went to interview. We were talking of the discussion that was being had at the time about whether more organs could be procured in America by giving people on death row the option to donate organs before they were electrocuted. He and I were both thinking about this with some horror, and, obviously, there are huge ethical questions about whether people would feel they were being coerced into doing this kind of thing. He quickly went on from there, but I could see there was more difficulty for him, and I said, well, what else is bothering you? He said, well, I wouldn't like the heart of a murderer in my body. Then he looked down at my tape recorder and realized that it was running, and he became a little concerned and tried to cover it up a bit. But he clearly felt that certain things are passed along through bodily organs — at some level. I mean, in his daily work he obviously didn't think that at all, but we all have these slightly mystical feelings, I suspect. Or very, very many of us. The transplant world is embroiled in these kinds of things, although they keep it damped down.

DC: Organ transplantation provokes powerful and contradictory feelings, as the story of the heart surgeon shows. But in Western countries, these feelings have been, as Lock says, damped down, while in Japan the issue has continued to rage. According to the most recent figures I could find, organs were taken from only forty-seven brain-dead donors in Japan last year. The comparable figure for Canada is close to two thousand and, for the US, twenty-five thousand. Margaret Lock has already cited some of the reasons for the difference: a less pronounced mind-body split in Japanese culture, the greater authority of the family, revulsion at the short-circuiting of natural reactions when the body of the beloved is snatched away and redistributed. But there was some initial revulsion in Western countries, too. Articles in both medical and popular journals worried that avidity for fresh organs would create an indecent, vulture-like interest in death. One writer evoked Nazi Germany's "suicide-assistance squads." Another imagined what he called "cadaver farms." So why did Western countries eventually embrace what the Japanese could not stomach? Here Lock draws tentatively on what might be called the deep cultural history of the West: the theme of resurrection, a history of medical dissection that stretches back to the Middle Ages, a habit of stifling natural reactions that interfere with technological progress. But, however the difference is to be explained, she says, the Japanese response is an arresting example of culture's continuing vitality in the face of techno-science.

ML: Cultural differences are still hugely important; what Veena Das has called "the persistence of the local" is not simply a sign of people's not being sophisticated, not being literate, not having exposure to advanced education. It is something much, much more profound and persistent than that, something in which the medical world itself participates. Willy-nilly, they participate. Many, many Japanese doctors would never want to procure organs. It's just absolutely counterintuitive to them, even though they know perfectly well that people will die if they don't get an organ. The Japanese have worked very hard on living transplants. They pioneered liver transplants in which you take a section of the liver from a compatible living person and give it to a child, and so on. They've been pioneers in this area and continue to work very hard at it because they feel that it's much more appropriate to take tissue from donors who are fully aware of what they're doing. This is a shining example of how cultural difference persists and pervades everything we do.

DC: *Twice Dead* deals, among other things, with cultural resistance to technological innovation. Lock's earlier work, also done in Japan, had been about the ways in which culture shapes nature. She'd gone to Japan first in 1964 because her husband was in training for that year's Tokyo Olympics as captain of the Oxford judo team, which represented Great Britain. Growing interested in Japanese culture, she ended up doing her PhD in anthropology on the revival of traditional medicine there. Then, later on, stimulated by the attention the subject was receiving in North America, she started to look at how Japanese women experience menopause. This would eventually lead to her book *Encounters with Aging: Mythologies of Menopause in Japan and North America*. One of the first things she discovered was that the current Western definition of menopause was far too limiting.

ML: Over the years, in the last fifty years, particularly in Europe, meno-pause has come to mean the end of menstruation. I mean, that's what the word literally says. But people of my mother's generation and earlier didn't really think of it just like that. They thought of it more as this midlife experience which was nasty and unpleasant, but they didn't focus nearly so much on menstruation per se and the symptoms that might arise as a result of its cessation, even though, obviously, many people experienced these symptoms.

When I went to Japan and started talking in Japanese about this, several things happened right away. First of all, in order to interview women in a factory, I had of course to approach the managers of the factory. They were men, and of course they all roared with laughter and said, why do you want to study this? Then they said, and why are you only studying women? I must have looked a little surprised because they said, men have *kônenki*, too, which is the Japanese word for menopause. I said, oh, really? Do they? Yes. We have bad times at midlife, too.

I sort of brushed that off, because obviously there was still something fundamentally different about this transition for men and women. But when I then went and started talking more seriously to women in order to develop an appropriate questionnaire, it became clear that for most of them, *kônenki* does not mean simply the end of menstruation. The end of menstruation is part of a larger concept for them which is much closer to the earlier approach in Europe. It is part of a change of life which also involves physical changes like deteriorating eyesight and hearing loss and greying hair and all of these kinds of things, or the beginnings of these things. There is a whole range of symptoms — things like shoulder stiffness, which is a very characteristic Japanese symptom, some kinds of special headaches, tingling in the hands and feet, a heavy feeling of the head, all sorts of special Japanese symptomatology — these are the kinds of things that people associated with *kônenki*. I would bring up the question of hot flashes, and then I found very quickly that this symptom that everybody in the West assumes is related specifically to the end of menstruation doesn't have a specific word in Japanese. There we have a language which is extraordinarily sensitive to changes in bodily states — so much so that Japanese women can look at me and say, oh, English is so dull, it doesn't have the variety of words to express these changes — and then, when I ask about hot flashes, I find I'm groping for language, because there isn't a single word that conveys the idea of a hot flash with the confined meaning that is expressed by the English term. That's a big signal to an anthropologist: this is not a symptom that is bothering most of these people. Otherwise they would have a clear, resounding word that unequivocally describes it.

In constructing the questionnaire, I learned very quickly that I had to use several different words to convey the idea of a hot flash in order to make sure that people reading it would understand. I also had to recognize that menopause does not translate into *kônenki* in any straightforward fashion. Another thing my colleagues and I found out very early on, when we started getting the first results from this questionnaire, was that a large number of women who had stopped menstruating for more than a year — in fact, a good twenty-five percent — said to us that they had no sign of *kônenki*. Yet they hadn't menstruated for a year, and these are women who are aged fifty to fifty-five. This sounds very odd, but it's another confirmation of the fact that the end of menstruation is not what they are focused on. So — a very different experience.

DC: The different symptoms experienced by Japanese women was one of Lock's most striking findings, and it was confirmed by her interviews with Japanese doctors as well as with women themselves.

ML: These doctors would say to me, when I asked them to list the symptoms of *kônenki*, they would say, shoulder stiffness. They would say, tingling in the hands and feet. They would say, headaches. They would say, maybe depression, maybe anxiety. And way down their list, some would say, oh, and perhaps *nobose*, or a hot flash-type experience. This was not the first symptom on their list, nor were night sweats. These are the two characteristic symptoms that we in the West associate with menopause, hot flashes and night sweats. Barely on the doctors' symptom list. These men are trained, go to international meetings, read international publications, and yet, when I'm asking them to think what their patients come and report, these symptoms are down towards the bottom of their list.

Now, this work was done in the 1980s, and of course things have changed a lot since then. Menopause has become much more medicalized in Japan. Japanese gynecologists used to make quite a lot of money out of doing abortions, and these are now not nearly so common because people are using contraceptives more effectively. Many of them are actually almost short of work. This is not true in tertiary health care, in hospitals, but it applies to the men who own their own clinics. They were looking around for how to drum up work, to some extent, and menopause was clearly an area where they thought they could improve their services. A good number of them started to show an interest in this part of the life cycle for the first time and really began to push for women to come for regular checkups at this stage, which hadn't been the case at all until that time. Over the years, there has been much more publicity, lots of media coverage. The idea of a hot flash has got pushed much further into public exposure, and people are thinking about it more. Doctors are talking about it more. But what is really interesting is that a young biological anthropologist, Melissa Melby, has been working in Japan for the past four or five years, and she took my original work and went and replicated it on a smaller scale. She finds that there is indeed increased symptom reporting, but it is still statistically way below the level of Canadian and American women's reporting. So something very interesting is still going on.

DC: How was this interesting difference to be explained? Medicalization didn't seem to fully account for it, since the difference persisted even as medicalization increased in Japan. Not only that, but the characteristic European symptomology had also been there before the doctors got involved.

ML: What happened originally in the West was that when the profession of gynecology first began to emerge, when interest in women's diseases, concern about women's diseases was transformed in the early part of the nineteenth century into the beginnings of the profession of gynecology, there were one or two doctors who were interested right from the beginning in menopause. The man who invented the term was a French doctor called Gardanne; he created the term "menopause." Before that, everybody had talked about the climacterium, but he deliberately invented this word in order to make it into a medical matter. He would get rid of all these diffuse symptoms that he thought were not appropriate; he would concentrate on the end of menstruation, what that meant, and how it needed monitoring by the medical profession.

Now, I haven't investigated other languages, but in English, prior to that time, there was indeed a term used among women: hot blooms. That obviously was the word that conveyed the idea of hot flashes before the medical world got anywhere near it. There's no equivalent that I can find in Japanese, although I've had Japanese readers looking for one from earlier in their history. This suggests to me that there was more of a concern or an interest or a sensitivity or sensibility about this kind of symptom, in England at least, and one assumes in many other parts of Europe and among the European inhabitants of North America.

DC: The more she reflected on the persistent differences between the physical experiences of Japanese and Western women, the more Lock began to wonder if these might be biological and not just cultural differences. But to think this way was to offend a central belief of her professional community, and indeed of all modern sciences. "Nature," as Spinoza had said, "is always the same, and everywhere one." And the body belonged to nature.

ML: It is anathema to most cultural anthropologists to even entertain the idea of recognizing biological difference. There has been, as everybody knows, this thing called the nature-nurture debate over the last hundred years, in one way or another, and anthropologists, if they're cultural anthropologists, if we're all shaken out, come down on the side of nurture. In other words, cultural anthropologists have tended to assume that there is a universal body, or it's near enough universal. In any case, it's not the realm of a cultural anthropologist to investigate, so we have black boxed the human body, left that to the biologists. We have gone off and investigated differences in different parts of the world, and differences within North America and Europe and so on, based on language, on expectations, on cultural ideas, on politics, on the different ways medical professionals go about doing things, and on alternative medical practices. All of these sorts of things have fascinated cultural anthropologists, particularly medical anthropologists.

For me to take a step into the world of thinking about biological difference — I knew that this was going to be fraught and potentially difficult to get my close colleagues to accept. But I felt that I had to do this. I didn't just do survey research in Japan. I talked to well over a hundred Japanese women, probably approaching two hundred Japanese women, about this, and also, as I've said, to doctors, and to feminists, and I was absolutely convinced that there was something really, really biologically different going on here. The difference was not just in the way things were being expressed. So I created this concept of local biologies, in which my position is that there is indeed significant biological variation among different human populations.

DC: In speaking of local biologies, Lock was suggesting that there is a much more lively interplay between biology and culture than either cultural anthropology or evolutionary biology had formerly supposed.

Biology is not a universal constant, she argued, but something that is actively shaped by the ways in which people live.

ML: Not only originally evolutionary changes, but historical and environmental factors have impinged on the actual physical body. Dietary practices, marital and reproductive regulations, these sort of things account for changes in the actual physical body, and these changes accumulate in populations that tend to intermarry a lot. They're not static. They're ongoing, and they vary, to some extent, from one part of the world to another. However, they do change. When Japanese come to live in Hawaii or in the States or Canada, then — no doubt due to differences in behaviours, differences in diet — there are major changes in their actual physical bodies. In other words, my basic hypothesis, which is shared by many, many people, is that the environment influences biology, that biology is not this static entity. It's mobile. It's changeable. And environments and social change have a profound impact on biology.

I argue that with many of these issues it helps to start to think about the biology and the culture as interrelated in a way which makes it very difficult to separate them one from the other. You can't just measure biological change or just talk to people. Neither of these by itself is going to be satisfactory. You have to integrate these as best you can to get a fuller and richer picture of what is going on.

DC: This new picture of biology and culture interwoven has received a lot of support since Lock first challenged the prevailing orthodoxy. Subsequent studies in China, India, Southeast Asia, and Mexico have confirmed the wide variation in symptoms that she found in her comparative study of menopause in Japan and North America. And biology has given birth to a new field called epigenetics, which studies the interaction of genes and environment. Research in this field has increasingly undermined received ideas about genes.

ML: A huge amount of the new biology, the new molecular biology, coming down the pipeline is beginning to show that the way in which we have understood genes and their function has profound limitations. It has helped us and provided many insights, but it has profound limitations. Indeed, it looks as though there is a strong possibility that things that happen during one's lifetime can indeed be transmitted to the next generation. Of course, they get transmitted through culture, i.e., dietary practices and so on, but there are also, we are now beginning to realize, some things which apparently can be transmitted through biology, too.

DC: Those familiar with the history of biology may recognize in what Margaret Lock says here the shadow of an old heresy known as Lamarckism. Writing around the turn of the nineteenth century, Jean Baptiste Lamarck put forward the first really coherent theory of evolution. However, he argued as part of it that acquired characteristics could be inherited, an idea that was later rejected, sometimes quite violently, by modern evolutionary biology, which held that our genetic constitution can vary only by chance, with the lucky chances providing an edge in the struggle for existence. Lamarck's view became a scandal in the eyes of orthodox evolutionists.

This was a highly polarized debate. Either the genetic material was separated by a firewall from all outside influences or it was substantially reshaped in each generation. Today, a more nuanced picture seems to be emerging, and it includes the idea that some acquired characteristics may be passed on. A remarkable study seems to support this formerly heretical idea.

ML: The study is from Holland, and it's to do with what is called the Dutch famine, which took place towards the end of World War II. The country was occupied by the Nazis, and there also had been a very bad year in terms of crops, the success of the harvest, and things like that, so a very large number of people in Holland were deprived of food. By some remarkable foresight on somebody's part, they kept records of the women who were pregnant at the time and tracked their babies when they were born. They found not only that the women themselves tended to succumb to late-onset diseases more than would have been predicted — they had higher rates of cardiovascular disease and other sorts of problems — but so did their children when they grew up. Now a third generation, the grandchildren, are beginning to show similar problems, signs of much too great an incidence of some of these late-onset diseases. The researchers link this to the deprivation that those pregnant women went through. They're hypothesizing that certain genes were not switched on at the right time or else their products were produced in excess to compensate for the environment, and that not only did this affect those women, but it has been passed on to the next generations.

Now, this is the beginnings of some of the really exciting research that's coming out in epigenetics, which I am not really competent enough to explain to you, but you get the sense of what is going on here. These will be epidemiological and basic science studies which give little glimpses into what is a much more complex picture that we need to pay attention to. In terms of teaching genetics, the first shift is to move to recognizing that genes always have to be activated. They have to be switched on, and then they get switched off at certain times. To focus on normal development, what happens in the process of normal development as far as genetics is concerned, would be much more helpful and much more appropriate than focusing on the kind of mutational genes that produce really unusual diseases. Obviously, those unusual diseases cause enormous suffering, and we need to keep working on them, but this is what's tended to happen a lot with medicine. You use pathological examples to illustrate instead of using normal growth and development as your illustrative example.

DC: Margaret Lock speaks here of genes acting only when they have first been, as she says, switched on, or expressed. This is a key idea of the new genetics. When the word "gene" was first coined by biologist Wilhelm Johannsen in 1909, the term was purely speculative. Heredity must have a cause, and that cause was called a gene. It was biology's version of the atom, the irreducible something that must be at the bottom of everything else. As late as the 1930s, there was no consensus among geneticists as to whether genes were real or, as one eminent geneticist said, purely fictitious. Then came the modelling of DNA, the master script in which everything is written. Genetic determinism reached a high point. Genes were tiny dictators, impervious to their environments, giving orders but never taking them.

Little or nothing is left of this picture today, as you've already heard. More recent research has shown genes to be, as one of the pioneers of molecular biology put it, marvellously communicative. They are not little dictators at all but exist in a dynamic interchange with the rest of the cell, giving and taking, sometimes expressed, sometimes not. These new findings dovetail very neatly with Lock's research and support her idea that people vary biologically as well as culturally.

Lock's concept of local biologies has important implications for the vast complex of biological science and medical practice that she calls biomedicine. Biomedicine, in her view, has too often assumed that its knowledge is universally valid when it is, in fact, quite partial. The gospel she has been preaching is recognition of diversity. Bodies vary from individual to individual and culture to culture, but biomedicine still tends to assume that they are everywhere the same. As a result, it obscures and suppresses variety.

ML: Biomedicine standardizes. It assumes that there is a universal body, that the results of the clinical trials that we do will be of use wherever we take them, and that our approaches and methodologies are applicable anywhere. This idea of standardization and of uniform practice is disseminated around the world, even in the many places where, because of poverty, the facilities are dreadful, the doctors can't do what they've been trained to do, and so on. Even there, you begin to get publics, patients whose ideas are being transformed quite rapidly — not in depth by biomedical knowledge, but by a new way of thinking about the body. Essentially the body is decontextualized from its social environment.

In traditional medical systems of all kinds, individuals are recognized, above all, as being part of a social milieu which, it is assumed, has a major impact on health, illness, and well-being. But biomedicine, for the most part, has shaken off that social context to deal with the body itself, though there are partial exceptions, like public health and some aspects of psychiatry. The assumption, on the whole, is that the body is a universal.

DC: The effect of this assumption, Lock says, is to force people to ignore their own experience. The terms on which they know themselves are no longer relevant to their treatment. Thus patients lose their voice and learn to address biomedicine in the terms it understands.

ML: The patient's story becomes irrelevant, because that's all established through measurements and through imaging and through standardized questions; the patient's own narrative account becomes useless. The patients themselves, when they go for HIV treatment or help with HIV, or when they go for help with their newborn infants — the whole range of issues that people take to medical professionals — begin to pick up the way in which they are supposed to understand their bodies by recognizing what is being done to them. No longer are they asked about their social life. They are asked, in detail, about their symptoms. People learn how to be patients. They learn how to report what they're supposed to report, and they learn what not to say. My colleague, Vinh-Kim Nguyen, who works in West Africa, would, I'm sure, be one of the first to say that in parts of Africa people have learned already that you certainly don't go and talk to the biomedical practitioner about things like witchcraft or other forms of social problems — those are inappropriate. You do talk about your symptoms, and you certainly hope that you're going to get treatment for those symptoms.

Many of the alternative medicines — Chinese and Japanese medicine, Ayurvedic medicine, and many of the complementary medicines, as they're called — have also taken on this sort of biomedical approach. They don't all necessarily accept the idea of the universal body in such a uniform way, but there is a sense that symptom reporting and attention to the internal workings of the body is absolutely crucial. The social dynamics and the social determinants of health and illness, as the epidemiologist would say, tend to get pushed to one side. A lot of the work that medical anthropologists have done is to look again and again and again at the way in which conditions that we would tend to assume are largely socially and culturally and politically produced then become individualized and medicalized. All the social aspects get dropped off and lost in the dust while one focuses on the body. Something like Attention Deficit Hyperactivity Disorder would be a current example of this kind of thing. Now, this is not at all to say there aren't some children with real problems, but to focus entirely on medication is to assume that all of this results from problems internal to the body and that whatever's going on in the environment doesn't affect the body that much.

DC: It's the job of medical anthropologists and other social scientists, Lock thinks, to bring a social perspective into biomedicine, and there is greater openness to social science perspectives within medicine than there used to be. The Department of Social Studies of Medicine to which she belongs is an example: it's part of the medical faculty at McGill University, but it is also connected to the departments of history, sociology, and anthropology. She sees the proper relation between social and medical science as one of dialogue. The local and the universal should be brought into conversation. Insisting that there are no general principles would be just as bad, from her point of view, as having nothing but general principles. Circumstances make cases.

ML: It seems to me that a broken leg is a broken leg, and that a good number of other medical conditions are essentially conditions that could be managed in exactly the same way wherever you are. This would apply to trauma — physical trauma — and some very fundamental ill-nesses, but not all the infectious diseases. We're finding with HIV and with TB that things are more complicated than we formerly thought. Sickle cell anemia doesn't seem to be a straightforward problem at all. People in different areas of Africa experience it in different ways and apparently have different symptomatologies. There's a lot of interesting things that we have to learn that we haven't looked for. We've damped all this down.

But having said that, of course there are many things for which bio-medicine can be introduced and used in very, very effective ways that have never before been possible. There's no argument about that; there are many instances in which the body should be understood as more or less uniform and standardizable. Anthropologists tend to work on margins and work on exceptions, but the kind of problems I've been talking about are something more than just unusual exceptions. This is a very large area which includes all the psychiatric illness and mental health problems, many of the life-cycle changes, and many of the chronic conditions. There you find vast differences which are indeed quite significant.

DC: Honouring these differences, in Lock's view, will require major revisions to our received wisdom. Science will have to become less preoccupied by sameness and regularity and more interested in variation, and old dichotomies between society and nature, culture and biology, will have to give way. Cultural anthropology, she says, has treated difference as a matter of culture only, leaving nature itself untouched. But now these very categories have come into question.

ML: The earlier story was one of cultural construction, including cultural construction of ideas about nature, ideas about the body. The assumption was that people think in different ways about these very fundamental things — about the world, about nature, and about the human body — and that it's important to record how this is brought about. And indeed, that is true. One needs to pay attention to the cultural construction of the human body, but one also needs to pay attention to the biological body and to recognize that this body that we, as anthropologists who grew up in a Western environment, had tended to assume was essentially universal — which is the dominant ideology of the biomedical system — in fact shows some really important and interesting variation. And we need to hook this new account up to the way in which people are experiencing and talking about their bodies.

Now, obviously medicine has always acknowledged variations to some extent. Any decent clinician knows perfectly well that people respond differently to medications, and he adjusts accordingly and asks you to come back and tell him the result and so on and so on. Everybody knows that. But their assumption is that this is all individual variation within a manageable range. It is equally important to recognize, I believe, that this kind of variation, although a lot of it happens at an individual level, also happens at the larger level of biological populations. These things now need to be brought together. It's a huge challenge because you have to sacrifice the language of science, to some extent, or be willing to modify it and use it in ways that are not tightly tied to the results of clinical trials or to evidence-based medicine. These kinds of scientific tools are not suitable when one is looking for variations rather than for similarities.

If You Can Spray Them,
Then They're Real

IAN HACKING

Francis Everitt and I wrote a paper sometime in the 1970s called "Which Comes First: Theory or Experiment?" That's the only paper that both of us have had systematically rejected, and the rejection was always, who cares about experiments? We sent it to several different kinds of journals: a popular science journal, a professional philosophy of science journal, a general physics journal. They all said, "Who cares about experiments?"

— Ian Hacking

Philosophers of science tended until quite recently to treat science as a mainly theoretical activity. Experiment — science's actual, often messy encounter with the world — was viewed as something secondary, a mere hand servant to theory. Popular understanding followed suit. Theories were what counted. One spoke of the theory of evolution, the theory of relativity, the Copernican theory, and so on. It was as thinkers and seers that the great scientists were lionized and glorified. But this attitude has recently begun to change. A new generation of historians and philosophers have made the practical, inventive side of science their focus. They've pointed out that science doesn't just think about the world, it makes the world and then remakes it. Science, for them, really is what the thinkers of the seventeenth century first called it: experimental philosophy.

One of the people who has been most influential in advancing this changed view is a scholar who is widely regarded as Canada's pre-eminent philosopher of science, Ian Hacking. In his book *Representing and Intervening*, he tells the story of a conversation he once had with a physicist friend of his. This friend was conducting experiments designed to detect that famous but elusive elementary particle known as the quark. I'll omit the technical details except to say that the procedure involved varying the electrical charge on a very cold metal ball. "And how is the charge on the ball altered?" Hacking asked. "Well," his friend replied, "at that stage we spray it with positrons to increase the charge or with electrons to decrease the charge." From that day forth, Hacking writes, "I've been a scientific realist. So far as I'm concerned, if you can spray them, then they're real."

This epigram, "If you can spray them, then they're real," gives a little of the flavour of Ian Hacking's approach to the philosophy of science. Positrons are real because we can produce them and do things with them, not because we can prove that they exist eternally in the mind of God. Science, for him, is a creative activity. It brings new things into existence. "Many experiments," he has written, "create phenomena that did not hitherto exist in some pure state." And what is true of the things we make is, in his view, equally true of our mental capacities. New ways of thinking also emerge over time and, in doing so, change the very terms on which the world appears to us.

Ian Hacking grew up in North Vancouver and, after undergraduate studies at the University of British Columbia, went to Cambridge on a Commonwealth Scholarship. There he became an analytical philosopher and began the work that would lead to his first book, a study of statistical reasoning. It was published in 1965, and the reception it received left a lifelong impression.

IAN HACKING

I guess I really started the habit of philosophers — with one predecessor who was at Cambridge, Richard Braithwaite — started the habit of philosophers of looking at what statisticians actually do and how they reason, and so my first book was called *Logic of Statistical Inference*. It was a very, very good experience for me because I was assistant professor at the University of British Columbia when it came out — and with all respect to my former university, in 1965 it was basically a nothing university in the world's view — but I got all these wonderful letters from all the people whose work I had been reading. These were the really serious thinkers about what I'd written about, and they sent me criticisms and questions and suggestions and alternate points of view — two- and three-page typed letters (this was in the good old days). These were people whom I thought of as gods — I was in my twenties, and I was nowhere. That was a wonderful introduction to serious intellectual life which paid no attention to disciplines. These were mathematical statisticians who were interested in what a young philosopher had to say, and, of course, I was interested in them. I think that had the very good effect on me that I've never been shy about talking to experts, most of whom are dying to talk to you — and nervous — and are delighted if somebody understands what they're doing and has questions and wants to know about it. I continue doing that to this day. At the moment, I'm interested in a very, very, very contemporary bit of physics about Bose-Einstein condensation. I find I'm very comfortable doing what most of my philosophical colleagues cannot do: walking into a lab, or writing an e-mail and saying, can I come see you?, and the experts say, yeah, sure. I don't know who you are, but if you're interested, I'll tell you, and then just listening and so forth. So that was a very good introduction to what I might call relationships with working, really smart scientists.

DAVID CAYLEY

This close attention to what scientists actually do made Ian Hacking one of the acknowledged pioneers in the new field of science studies. He also explored new paths in the historical study of scientific ideas in a work called *The Emergence of Probability*. Published in 1975, the book explores a fateful innovation: the simultaneous appearance all over Western Europe around the year 1650 of a new way of thinking and a new kind of knowledge. "Probability" up to this time had referred to a thing's standing, not, as today, its likelihood. A well-regarded doctor was a "probable doctor," a comfortable situation a "probable way of living." Then in the years around 1650, the word took on the meaning familiar to us today: the degree of belief warranted by evidence or the way in which chance events fall into predictable patterns. So, for example, we might be able to successfully predict the number of accidents that will take place at a certain intersection every year, but each individual accident will apparently happen by pure chance. This was a radical break, a move from a world in which the standard demanded of knowledge was absolute certainty into a world where mere likelihood was the best that could be expected. Yet Hacking believes that this break itself was a chance event, the coming together of a series of disconnected lines of development.

IH: It's a complete accident that, because of the plague and certain other things, the city of London started producing what they called Bills of Mortality: every week, they would post the number of people who died of various causes in each parish within the city limits. Now, that's a total accident, but all these numbers are suddenly there. They were literally bills that were posted on the church wall, so you could actually see how the relative frequencies of deaths were changing. That's an accident.

There were lots of other accidents. Why was it that Leibniz, who was fascinated by the structure of games, should see the analogy between games and certain legal proceedings? It's an accident that the Dutch figured out that annuities were a good way to raise money. There are a million ways, as we know from history, in which governments get money out of their citizens. Well, the Dutch did it by selling annuities. And so on.

DC: "Annuity" here means that the Dutch citizenry invested in the state in exchange for a regular return. The fact that the Dutch government was financed in this way, rather than some other, Hacking says, had nothing to do with the fact that the English published Bills of Mortality. But the Dutch practice forced them to calculate probabilities in order to have sufficient funds to pay the annuities, just as the English practice made visible the fact that a population has a measurable rate of mortality. These coincidences, along with many others, amounted at a certain moment to a sea change in Western Europe's habits of thought, a change that has continued and accelerated into our own risk-obsessed age. How probabilistic thinking eventually came to dominate modern societies was a story Hacking would tell in a second book on the subject called *The Taming of Chance*. This book recounts what happened in the nineteenth century, when Western Europe was suddenly inundated by what he calls an "avalanche of printed numbers."

IH: The French government starts publishing the annual rates of crimes and suicides and prostitution and whatnot, and they're all tabulated and so forth; it's done for bureaucratic reasons, and it has to do with the change in the structure of society. In the eighteenth century, all that data was totally hidden. It was a state secret. But then these numbers become generally available. People start realizing that there are a large number of regularities in social phenomena and gradually come to think of things that happen in the world in a probabilistic way. One of the things they got all excited about, for instance, was the fact that the number of suicides in each arrondissement, each little administrative region, in Paris was the same, give or take two or three people, every year. Every year, in each tiny little district, the same number of people commit suicide. They thought, gawd, here we have something which is totally aleatory, something that is totally random, yet which is covered by laws. That's why I called my book *The Taming of Chance*. There's this curious mixture. Chance is something utterly undetermined, and yet it's subject to very general laws of a social or physical character. That really made people feel and experience the world in which they lived in a totally different way, and today it's institutionalized in every aspect of our lives.

DC: The habit of thinking statistically changes our very idea of ourselves, Hacking says. An example, taken from *The Taming of Chance*, is the idea of "normal." "Normal" is a statistical idea. The famous bell curve that guides the grading of examinations is what is called a "normal distribution." But "normal," when it slides from statistics into society, takes on an ethical colour. One wants to be normal. Statistical and scientific categories are never merely neutral. People begin to adapt themselves to the categories in which they are counted and described. Hacking calls this process by which expert knowledge feeds back into society "making up people." It's an example of how science not only observes, but also shapes the reality in which we live.

"Making up people" is the province of the social sciences. The case is not quite the same with the natural sciences. Quantum mechanics does suggest that, in some instances, observation can influence what is observed, but it can hardly be said that electrons learn to behave in the way physicists describe them. Nevertheless, there is a sense in which the natural sciences also change the world. This was a subject that Hacking took up in his book *Representing and Intervening.* One of its starting points was an ongoing conversation with a physicist called Francis Everitt, a colleague at Stanford, where Ian Hacking taught in the 1970s.

IH: Francis Everitt and I wrote a paper sometime in the 1970s called, "Which Comes First: Theory or Experiment?" That's the only paper that both of us have had systematically rejected, and the rejection was always, who cares about experiments? We sent it to several different kinds of journals: a popular science journal, a professional philosophy of science journal, a general physics journal. They all said, "Who cares about experiments?"

The real message of *Representing and Intervening* is that much of science is experimentation and changing the world and building instruments to change the world — that's the intervening — and not just theorizing, or representing. John Dewey spoke very critically of what he called the "spectator theory of knowledge" — the idea that all our knowledge about the world is got by looking and thinking — but many philosophers of science have disagreed. The most influential philosopher of science in those days was Karl Popper — I don't mean for philosophers, but for scientists and the general public — and I have enormous, enormous respect for Karl Popper, more than many of my colleagues — but Popper said all real science is theorizing. The experimenter is just there to provide tests of the theories that theorists have put forward. Before the experimenter can begin, the theorist must have done his work. So theory was what fascinated philosophers, but it also fascinated the general public, right across the board, I would say. One of the things which I tried to do in that book was to re-introduce serious reflection on experimentation.

DC: Hacking wanted to make experiment the equal of theory in both the philosophical and the popular image of science, and he was able to show in his book that experiment is something much more than a mere test of theory. Experimental findings sometimes run far ahead of theory. The phenomenon of Brownian motion, for example, was discovered eighty years before Einstein was able to explain it. More than that, Hacking says, experiment actually creates new things.

IH: What experimenters are doing is creating an instrument to interfere with the world, to intervene in the world, and that, I think, is the discovery: that we can really make instruments that change the world and produce — create — new phenomena. There's a chapter in that book called "The Creation of Phenomena" about a new ability of ourselves that we are still only beginning to understand fully. We induce physical changes in the world and create phenomena that don't exist until we have made them. Yes, we make them in the light of theory, but then theories have to be constantly remoulded to fit with the phenomena that we discover.

DC: *Representing and Intervening* was published in 1983, the first of a number of books that would give a new prominence to the productive role of experiment in science. But it wasn't just the academic account of science that was changing. The relationship between theory and experiment in scientific practice was also changing at the same time. Hacking found this out recently when, as a professor at the Collège de France in Paris, he resumed his direct study of physics.

IH: Until three years ago, I had not thought seriously about experimental science since publishing *Representing and Intervening*, but then I got interested again — for totally accidental reasons — while I was at the Collège de France. I thought I should be learning something new and current, and there was a one-year professor at the Collège de France,

Sandro Stringari, who is a theorist and phenomenologist of what's called Bose-Einstein condensation, which is what happens when something gets very, very, very, very, very cold — to within a nano-Kelvin, a ten to the minus-ninth of a degree centigrade, above zero.

DC: "Zero" here — just to clarify — means absolute zero, the point at which classical physics believed all motion would cease, but near which contemporary physics is finding astonishing new phenomena.

IH: I found this fascinating. Stringari was not an experimenter, but I thought, if I'm going to find out about this, I should go to some labs — I mentioned at the beginning of our conversation my habit of visiting laboratories. I went to a number of laboratories and have become very interested in what happens in this field.

One thing which has become clear to me is that the division of labour between theoreticians and experimenters, which was still fairly commonplace thirty or forty years ago, is now much less sharp. The first really major lab that I visited was in Innsbruck, in Austria, and then later on in the year, I went to Boulder, Colorado, which was the first laboratory to actually demonstrate the phenomena of Bose-Einstein condensation, for which they shared the Nobel Prize. I also had access to the lab in Paris where they had won an earlier Nobel Prize for developing the techniques that had been used in Boulder. My colleagues at the Collège de France were essential for that. You can see that I was well placed. What was remarkable about the Innsbruck and the Boulder labs was that in both places they said exactly the same thing: it's so great that we've got a really good theoretical group here who know exactly what we're doing.

Now, laboratories are always in the basement for the practical reason that you don't want any vibrations. You don't want to have a ninth-floor laboratory because it's going to wave in the wind, and everything's going to be messed up. Therefore, you put it down in the basement, on solid foundations; then you have all sorts of other things to stop it vibrating as well. So the lab is in the basement, even if the experimenters have offices somewhere else. The scientists said, it's so great. We've got a group of people up there on the fifth floor who actually know what we're doing. Then I went up to talk to the group on the fifth floor, and they said, it's so great. We've got a group down in the basement who actually know what we're doing. These are different cognitive skills, you might say. The ability to get an experiment working is very different from the ability to articulate a theory. Some people can do both. Everybody has to do a little of both, but we're born with different talents. You're a broadcaster. I'm a philosopher. You'd probably be a bad philosopher, and I'd be a terrible broadcaster, so we have different talents. The important thing which struck me about this work and which seems to be substantially generalizable — at least in this realm of physical science — is that the picture of the experimenter and the theorist existing in different worlds and belonging to different social classes in many ways is becoming obsolete. I think it's because the whole field of at least physical science has evolved during my lifetime.

DC: Ian Hacking recognizes that his Bose-Einstein studies involve a branch of physics in which laboratories are small and that this fosters closer relations between experimentalists and theoreticians than could exist in the big particle accelerators and colliders, where thousands of people might work. Even so, Hacking believes that theory and experiment generally have been drawn into a closer, more responsive relationship in our time, and he predicts a dramatic consequence. Earlier philosophers of science, like Thomas Kuhn, Karl Popper, and Gaston Bachelard emphasized revolutionary breaks in scientific thought, but the future may be very different.

IH: There's a good reason why people were so fascinated by scientific revolutions — Popper and Kuhn alike. There have been really amazing, amazing changes, not just in the theory of relativity, which everybody knows about, but also in quantum mechanics. Think of what was supposed to be absolutely certain in Kant's time, absolute space, absolute time, absolute causality — all in shreds. The whole world was rebuilt early in the twentieth century. Those thinkers, like Kuhn and Bachelard and Popper and so forth, emphasized those kinds of radical breaks and radical changes in theory, whereas now I don't think there are going to be — this is a strong statement and one that I'd be happy to see refuted — I don't think there are going to be any more revolutions in physics. And I think it's because of the way in which the theorists and the experimenters have somehow discovered a harmonious way in which to work with the world.

DC: In his book *Representing and Intervening*, Hacking argues that, before the invention of modern science, thinking about the world and intervening in the world belonged, as it were, to different departments. However, "natural science since the seventeenth century has been," as he puts it, "the adventure of the interlocking of representing and intervening." Reality has, so to say, been put to the test. Unthinkable things have been produced — the way things are at one nano-Kelvin above absolute zero, let's say. But philosophy lagged behind. It continued to insist that representation is the royal road to reality. The experimenters remained hidden in the shadows of the theorists. Twenty-four years ago, Hacking called for a change: "It is time," he wrote, "to recognize science for what it is: the meshing of thought and action, theory and experiment." Today, he says, we have a truer picture of science and science, a truer picture of itself.

The Dance of Agency

ANDREW PICKERING

We don't live in a world of ideas. We live in a world of performances. We should think of ourselves as animals that do things in a world that does things. The world is very active. It doesn't sit around, as traditional philosophy would like to think, just being there and waiting to be described. It does things

— Andrew Pickering

In his book *Representing and Intervening*, Ian Hacking points out that science doesn't just theorize about the world, it changes it. At the time he wrote, a number of other scholars were thinking along very similar lines. One of them was Andrew Pickering, who today makes his academic home at the University of Exeter. Like Hacking, he was looking for a description of science that could take account of how scientific knowledge is actually made. This quest eventually brought him to think of scientific knowledge not as a picture of the world, but rather as a kind of performance. A year after *Representing and Intervening* appeared, in 1983, he published *Constructing Quarks*. The book is a study of particle physics and of how what he calls the "old physics" that had prevailed in the 1940s and 1950s gave way in the 1960s and 1970s to what he calls the "new physics."

ANDREW PICKERING

What struck me about these two different stories about the world of elementary particles is that they were almost disconnected from one another. They involved a different bunch of theories. Interestingly, they involved different kinds of machines. People built different kinds of accelerators or they used colliders. So there were different kinds of experimental setups that studied different phenomena and used different ways of processing data. The two worlds were, as I put it, incommensurable. That's a term from Thomas Kuhn's *Structure of Scientific Revolutions*. They just didn't map onto one another, which meant that the history of physics wasn't a continuous process of the accumulation of knowledge. It was a discontinuous shift from one way of understanding doing physics to another.

DAVID CAYLEY

Pickering was well qualified to study the world of particle physics. He had done a PhD in physics at the University of London, gone on to post-doctoral work at the Niels Bohr Institute, in Copenhagen, and was working at the British high energy physics lab at Daresbury when he decided to take another direction. He moved from science to science studies, studying and working at one of this new field's most influential beachheads, the Science Studies Unit at the University of Edinburgh. And he engaged with the thinker who was then on everybody's mind, Thomas Kuhn, the author of *The Structure of Scientific Revolutions*, a book that treated the history of science not as a continuous progress, but as a discontinuous succession of what Kuhn called "paradigms."

AP: Before I ever went to Edinburgh to start working on science studies, I read Kuhn's *Structure of Scientific Revolutions* in the University of London library one afternoon, and early on, I was very annoyed with it. I thought, this picture just doesn't seem right. One of the first things I did when I went to Edinburgh was to give a little seminar to the members of the group there on what I thought about Kuhn, and also on Feyerabend's philosophy and maybe on the thoughts of some other people. I can remember, at the end, I held up *The Structure of Scientific Revolutions* and dropped it in the wastebasket. I felt that I'd disposed of it. Needless to say, I was totally wrong. There's an enormous amount of good stuff in *The Structure of Scientific Revolutions*. One great thing about Kuhn's work, especially that book, is that it stresses the dynamic aspects of scientific practice. I don't think you get anywhere in thinking about science if you try to understand it on the model of truth. Scientists do not sit around and say, is this theory true? They sit around and say, what am I going to do next? What's my next project? What's my next experiment? What can I do with this theory? Where's my next grant coming from? is also entangled with that. Kuhn had a very nice image of what I ended up calling the dynamics of practice. This is his idea of the paradigm as an exemplar or model. A paradigm is some successful piece of work that you can model your future practice on, which means doing something like it, but different. Science grows in this process of modelling, and that's something that I definitely took from Kuhn. The other idea that I took from Kuhn was incommensurability, that somehow different scientific communities with different paradigms live in different worlds. I was very interested to find that you could make exactly the same kind of argument about the history of particle physics.

DC: In his book *Constructing Quarks*, Pickering shows the thought-styles that he names the "old" and "new" physics as self-contained worlds, each distinct and, as he says, "disjoint" from the other. He also gave what was then a new importance to the role that experiment plays in constituting these different worlds, each world depending on which experiments are chosen, which apparatus is used, and how the results are interpreted.

AP: When I first got into science studies, experiment was an almost totally neglected category. The philosophy of science and the history of science had largely been preoccupied with theory. The history of science was, in that sense, largely a history of ideas. What I got from my early controversy studies was that the materiality of science — experiment, instrumentation, machines — is really important. This was a surprise to me because I had been a theoretical physicist myself. Even so, I realized that the production of knowledge largely depended upon struggling with experimental setups and instruments and machines. They were recalcitrant things. You couldn't make them do whatever you wanted them to do. In a sense, you were at the mercy of your experimental setup. From the start, I was convinced that these kinds of material struggles should be very much a part of our story of science in history and philosophy and sociology. They somehow had to be brought into the discourse of science studies. A lot of my work is focused actually upon doing experiments: the production of facts and then how facts relate to higher conceptual structures.

DC: In writing about science as something produced or constructed, Pickering was putting forward a view that was, at the time, controversial; it would soon lead to what became known as the science wars. Critics of his approach insisted that science gives us the world as it is, not as we imagine it to be. For them, to say something is constructed is to diminish its reality, but Pickering says that he has never doubted that the world is real.

AP: I don't think reality is a problem. Reality is just there. We're in it. We have to deal with it. We don't live in a world of ideas. We live in a world of performances — that's the word I've ended up using. We should think of ourselves as animals that perform, that do things, that are always trying to do new things in a world that does things. The world is very active. It doesn't sit around, as traditional philosophy would like to think, just being there and waiting to be described. It does things. We had a tremendous thunderstorm here last night. The world was very definitely doing things. One of the phrases that I came up with in my later book *The Mangle of Practice* was the "dance of agency." We try to do things. The world does things back. We respond to that. The world responds to us in a kind of open-ended, never-ending, emergent process. That's how it is. That's how I claimed it was in science in my book *The Mangle of Practice*. I find it very interesting to try and recognize that fact. Science, the modern sciences, physics especially, depend upon not recognizing that fact.

DC: Physics fails to recognize the dance of agency, according to Pickering, because it remains fundamentally Platonic. It holds that behind the flux of appearances stand eternal and unchanging ideas. But Pickering's studies gradually convinced him that scientific knowledge is much more a matter of a give-and-take, in which the world is as active as the one who aspires to know it.

AP: I started thinking, well, maybe the world really is like that. Maybe it is full of performances, going back and forth between people and people, people and things, people and knowledge. That general image was the one that I sat down and tried to write out in my book *The Mangle of Practice*, which came out in 1995. So the mangle was, in those days, the phrase that I was using for this difficult back and forth that you can see in the scientific laboratory, but you can also see it in the workplace, and you can also see it in conceptual practice.

DC: Why "mangle"?

AP: Originally, I talked about "the dialectics of resistance and accommodation." I said, scientists in the laboratory usually have some goal in view. They imagine producing a piece of apparatus which will do X, which will help them answer question Y. What they always and inevitably find is that, when they build the apparatus, it doesn't do what it's supposed to do, and I called that process the "emergence of resistance": the goal and the thing itself don't fit together. I said, how do we respond to that? We accommodate ourselves to it. Maybe we think of a new goal, or maybe we fiddle around with the apparatus and try and get it to do what we want it to do. That's our accommodation. And then we go back to the world and see what it does, backwards and forwards. That's what I meant by a dialectic of resistance and accommodation. I think that's a good description of how science evolves.

But try saying "dialectic of resistance and accommodation" over and over again. It's one of those . . .

DC: . . . I know, I'm a broadcaster . . .

AP: . . . right, and then try writing it down again and again and using it as a verb and then in an adjectival form and things like that. I thought, I need a different word. I thought, I'm going to call it the "mangle," and that's good because you can use "mangle" as a noun or a verb. Of course, it's also the name of the bit of equipment that women, mainly, used in the kitchen to squeeze water out of the washing, and it amused me to appropriate the name of a bit of kitchen equipment to describe the history of high-status sciences. There was a kind of dissonance there. But it's got the idea of unpredictable transformation in it, and you can also say things like, I was mangled in a traffic accident, so it has a violent edge as well. It's an interesting word, but it's not a great metaphor — I will concede that.

DC: Andrew Pickering began his search for an adequate image of scientific practice during his time at the Science Studies Unit at the University of Edinburgh. It was home to what his colleague there, David Bloor, called the "strong program" in the sociology of knowledge. The strong program claimed, to put it briefly, that science could not itself escape scientific explanation, as if it were an eye that sees but is never seen. This explanation must be something other than science's account of itself; it must come from outside science. It followed that the cause of science could lie only in the surrounding society. Scientific explanation of science was social explanation. Pickering says he learned something from this approach, but eventually he concluded that it was virtually a mirror image of the kind of scientific thinking that he was trying to get past, scientific fundamentalism replaced by sociological fundamentalism.

AP: We discover a different hidden reality, so instead of the quarks being the hidden reality, it's something like social interests or social structure which is the hidden, stable reality, the reality that explains the flux of appearances. There's a kind of inversion strategy here. Instead of finding the hidden structure in matter, you find it in the social. This all has to do with the structure of the modern disciplines: the physicist's job is to find the hidden structure in matter, the sociologist's job is to find hidden structure in society. The academic disciplines give us the apparatus for finding these hidden structures wherever we want to find them. That's how the traditional disciplines reproduce themselves. That everything can be reduced to some more basic explanation is modern common sense. When we watch the news, we watch for sociological explanations. What are the interests behind what's happening in Washington today? This is a very natural way of talking. That's why the science warriors could get to grips with the sociological position: it was another familiar position. They just wanted to say, it's wrong. What my studies of the history of science have convinced me of is that this kind of partition just doesn't work. It's not that there are many stable realities in different places and that we can add them up. It's that all of these different aspects of reality — matter, the social — are themselves in flux in relation to one another.

DC: In his attempt to describe a world in flux, Pickering has recently been drawn to the field of cybernetics as it took shape in Britain after World War II; his new book is called *Sketches of Another Future: Cybernetics in Britain 1940 to 2000*. Among the British cyberneticians of this period, the best remembered are probably Gregory Bateson and R.D. Laing. What interested Pickering was the style of thought he found amongst these maverick intellectuals, a style of thought very close to the one he had been trying to develop.

AP: The term comes from the Greek word *kybernetes*, meaning "steersman." In a sense then, cybernetics means the science of steersmanship. Then the question is, what does that mean? I think it means something along these lines: Imagine sailing a yacht, a boat that's driven by the wind. How do you do that? Do you do it by thinking about what's going on and calculating the forces of the wind and the waves and making precise adjustments to the tiller and so forth? No, no, you don't. You'd be completely stuck if you tried to sit there doing little calculations about how to sail a yacht in the wind. Somehow, that kind of sailing, that kind of steersmanship is the art of performatively getting along with a world that you don't cognitively understand. You lean on the tiller a bit more. You pull in the sails a bit or you let them out. You find a way of getting along in a very active world without being able to fully know and control how that world works. So I think — and this is certainly true of the history of British cybernetics — that cybernetics was a science — and much more than a science — of getting along in a world that's always going to surprise you, that you can never dominate, that will often offer you things that you don't expect in advance.

DC: Discovering the field of cybernetics surprised Pickering. In his 1995 book *The Mangle of Practice*, he had tried to put forward what he called a "performative theory of knowledge," a theory in which knowledge doesn't just float ethereally in the brain; it actually bumps up against the world. But he had not known at the time that people had actually made working models of this kind of knowledge-in-action many years before.

AP: When I wrote *The Mangle of Practice*, I thought of my argument as being purely theoretical. This is the best kind of description I can give of the history of science. When I started looking at fields like cybernetics, I realized that people had found ways of playing out these theoretical ideas. The cyberneticians had much more imagination than I had. They made up these wonderful projects and carried them through.

DC: What would be an example or two of that playing out?

AP: There are many examples. I'm not really sure where to start. The cyberneticians started by thinking about the brain. It's a not-very-well-known fact that cybernetics was largely brain science when it started, and my cyberneticians built very interesting model brains, electromechanical model brains. Their specific take on the brain, which is a very mangle-ish one, is that the brain is the organ of adaptation. It helps us perform and get along in a world that often surprises us. The important contrast here is with the usual idea of the brain as the organ of cognition and representation, the brain as being full of ideas. The idea of the brain as being full of ideas and thoughts and representations goes very nicely with the traditional history and philosophy of science, which treats science as being representation, ideas, et cetera. The performative brain that the cyberneticians were interested in goes very nicely with the mangle, which emphasizes performance, finding out, adapting to the unknown. In 1948, one of my cyberneticians, Grey Walter, built beautiful little robots that he called "tortoises," or "turtles," they call them in America, little robots which would explore their environment and find their way around in it, go up to lights, circle around them, wander off and look for more lights, avoid obstacles — things like that. This was back in 1948. Walter made extremely clever robots which he also considered to be models of the performative brain.

DC: Grey Walter's tortoises gave Pickering an image of how our minds, in effect, feel their way around. He doesn't at all deny that the mind also represents the things of the world and that we steer by these representations as well. He just thinks that we need to balance these two modes instead of claiming that representation is all of thinking. To this end, he proposes what he calls a "new ontology," a word that comes from the Greek root for "being." An ontology is an account of what the world is, and the one that has prevailed in modern sciences, Pickering says, simply can't do justice to an endlessly surprising world.

AP: In a conventional education, we learn about the world as a finite, knowable place. You can read about it in a book. You can understand it. The whole thrust of modern engineering is that, having understood it, you can reconfigure it and bend it to your will. The modern ontology, since the scientific revolution and the Enlightenment, is an idea of the world as a knowable, controllable, dominatable place.

The cybernetic ontology is a vision of the world as a space of quasi-biological becoming, evolution, transformation, emergence, a place in which we necessarily get on by finding out, by trying things. I try and do something in the world, and something happens, and I respond to that. It's not primarily a cognitive place. It's not dominatable. It's a place that you find out about rather than control. So I use the word "ontology" to try and conjure up this contrast between two visions of how the world is.

DC: The British cyberneticians, as Pickering sees them, had a vision of a world of surprises, and they did science in an ardent, amateur spirit that he found equally attractive. Here, he borrows a distinction from two French thinkers, Gilles Deleuze and Felix Guattari, who contrast what they call the "royal sciences" with the "nomad sciences." Royal sciences reinforce the existing order. Nomad sciences open up new paths, and British cybernetics, in Pickering's view, was definitely a nomad science. That's why he calls his book *Sketches of Another Future*. For him, the British cyberneticians demonstrated the possibility of another path for science, even if it is still largely a path not taken. They showed that one can think differently, and, by doing so, they created a way of "putting modernity in its place."

AP: By "modernity," I mean this ontology of the world as knowable and dominatable. That dominates our imagination. That's the default way of thinking about being in the world. It brings with it a certain power and certain historical achievements. But I want to say that that's just one way of understanding the world. I want to say that there's this other way of understanding the world. With a different ontology, the world appears as a place of becoming, of revealing, of open-endedness, et cetera. I think it would be useful if we had two ontological repertoires, rather than one, so at least we wouldn't think we have to dominate everything and everyone all the time. We could think of operating as the cyberneticians did. The cyberneticians give us a lot of models.

I also want to say that the second ontology is more fundamental than the first, that I think it is a general description of how the world is, and the modern ontology is just one particular stance that you can take within that sort of a world. If we saw that, that would be a way of putting the sciences and engineering and all sorts of political initiatives in their place. They are just one way of trying to go on — rather risky ways, very often, often inviting disaster. Maybe we should just resort to that way of going on when we have to, and we should think twice before we do that kind of thing. That's the sense of "putting modernity in its place" that I was after.

DC: Putting modernity in its place, for Andrew Pickering, means seeing it as an option rather than as a destiny. The sciences characteristic of modernity, he says, aimed at domination, but now we are surrounded by the perverse and unexpected consequences of this project, among them, at the very largest scale, a changing climate. He proposes a new science that would be more tactful, more tentative, and more self-aware.

Risk Society

ULRICH BECK

I sometimes feel pity for the scientists because they are in a situation in which they have to ask for money and make claims that everything is certain and that everything will be done in a proper way. But in fact, they often don't know what the consequences of their actions will be, and they cannot know it before they do it.

— Ulrich Beck

Few people ever apply a name that sticks to an entire social order, but sociologist Ulrich Beck is one of them. In 1986, in Germany, he published *Risk Society*, and the name has become a touchstone in contemporary sociology. Among the attributes of risk society is the one he mentions above. Science has become so powerful that it can neither predict nor control its effects. It generates risks too vast to calculate. In the era of nuclear fission, genetic engineering, and a changing climate, society itself has become a scientific laboratory.

There was a time, not so very long ago, when people could still speak unselfconsciously about modern society. Some societies were modern, others were modernizing, on the way to modernity. A modern society was run on rational lines. It was subject to planning, prediction, and control. It developed, progressing to ever higher levels of prosperity and enlightenment. And it lived in the clear light of science. Such a society never entirely existed, of course, but for several centuries it was a guiding ideal. Today, that ideal is in tatters, and what helped to undo it was the very thing that had defined it: technological science. Rationality began to generate results that exceeded reason's capacity to comprehend or manage. Robert Oppenheimer contemplated this paradox in the desert at Alamogordo, New Mexico, at the site of the first atomic bomb test, when he saw a product of advanced mathematical physics rise up before him as a destroyer god. Ulrich Beck contemplated it, too, as he became aware, during the 1960s, of the incomprehensible magnitude of the threats that even ostensibly peaceful technologies were beginning to produce.

Today, Beck is the director of the Institute of Sociology at the Ludwig Maximilians University in Munich. I interviewed him in his office there, and he told me that even as student he had begun to be aware that modernity had outstripped the concepts with which sociology was trying to grasp it.

ULRICH BECK

Risk society starts when public opinion begins to recognize that the dangers that are produced by modernization are not being addressed by the institutions of modernity. This is what risk society is about. On the one hand, we are producing ecological endangerments, and maybe even destroying nature, through all kinds of processes, some unknown and unseen because we don't recognize them. On the other hand, we have institutions that try to respond to these threats, but those institutions don't fit the basic endangerments which we are producing . . .

DAVID CAYLEY

. . . don't fit. You mean, aren't adequate?

UB: Yes, aren't adequate There are many examples of this. One happened here in Munich. There was smoke coming out of local factories that looked very dangerous. A nearby village was nearly destroyed by the smoke. You could even see that industry was causing the endangerment of the people in the village. They went to court and tried to say, stop this, otherwise you are endangering us. The court said, well, of course we see this relationship, but because there is more than one factory — there are actually hundreds of factories — we cannot identify a single cause, and a clear and singular cause is the only principle under which we can draw a judgment which allows us to say, you're guilty.

So here you find a paradox. The more numerous the industries that are endangering people in other villages and cities, the less it is possible to make them accountable for what they do, and the less our system of law is able to solve this problem.

Another example is the accident that happened in the power plant at Chernobyl. Of course, Germany, and all of Europe, was very much affected. But, according to the norms the German bureaucracy had to obey, a state of emergency could be declared only if the accident had happened at a power plant in Germany; outside of Germany, there could be no officially recognized catastrophe. They didn't know what to do. Nation states organize their security on a state-by-state basis. They were ready to act if something had happened in Germany, but, when it happened in Chernobyl, they were not ready. They didn't even know if they were allowed to declare a state of emergency or to initiate the political actions that are the consequences of such a declaration.

The nation state is the source of the institutions that respond to possible dangers and catastrophes, but if you have consequences which go beyond the state's borders, there is no answer. And the more you look into details, the more you find the same situation. Our institutions are based on norms that do not reflect and cannot respond to the transnational or even global challenges which are being produced by radicalized modernity.

DC: Radicalized modernity is modernity that creates consequences it is unable to address. It's a victim, in a sense, of it own success. Science is very much part of this fate. It, too, undermines itself. Science's claim on power and privilege rests on the promise of certainty, the promise that it will give us a clear and unambiguous picture of how things are. Yet, the better science gets, the more it creates uncertainty.

UB: There is a huge controversy about the consequences of science. Experts differ. Whenever you come across a question of risk, you'll find all kinds of different expertise. There is not one expertise. There are many expertises. This is not a sign that the expertise is bad. On the contrary, the fact that there are many expertises shows that it is good expertise. There can always be a lot of different and conflicting expertise on risk issues. This multiplication of points of view within science, of course, undermines the monopoly claim of rationality, which is part of the progress of science. The scientists themselves, by concentrating on risk and risk analysis, undermine their own monopoly.

DC: The classical image of science as verifiable knowledge rested on the observation of simple physical systems. The complex and chaotic systems that science contends with today will always be subject to multiple interpretations. Risk issues, Beck says, are always contestable, and this has meant that science has had to cede some of its claim on truth.

UB: From a sociological point of view, science was the institution which took the position of the church and the priest. In earlier times, it was the church and the priest that would try to create security. When religions ceased to occupy the centre of society, then science took over, and our belief in progress is still, in some ways, a religious belief, with the scientists as the priests. This is still a very powerful image, but the more science has reflected on its own role in society — and this is true in many different disciplines — the more scientists themselves have produced doubts about science's claim to be a source of rationality and security and truth. I think that for quite a while science has been losing ground as a symbol and an institution which creates quasi-religious certainty about how to manage the world.

This recognition that science is not actually the institution to solve all our problems reflects what we might call the secularization of science, and it's a very necessary step in modernity. Decisions formerly made by experts and scientists are once again becoming political issues that have to be decided upon in public.

DC: Questions that used to be decided by experts have now returned to the political arena. Beck has cited two of the main reasons for this: the scale of risk and the proliferation of sometimes contradictory expertise. A third is the way that techno-scientific innovation has begun to escape experimental control. The only way to test the effect of atomic explosions on human beings was to drop bombs on Hiroshima and Nagasaki. The effect of more than six hundred million automobile exhaust pipes on the earth's atmosphere is not something that could have been tried in advance of building the cars. Contemporary society *is* the experiment, Beck says.

UB: You can maybe test certain elements of a power plant in advance, but to test the system as a whole, you have to build it. In many fields now, the logic is actually the other way around from what we used to suppose. In genetics, as in many other areas of technology and science, we have to make it, we have to produce it in order to test it. Actually, society is becoming a laboratory. The whole world is becoming a testing place for technologies of all kinds because they can no longer be tested in a separate space or container. Here again, we need a different methodology and some kind of public participation in assessing the results of this new kind of experiments.

DC: So you're saying that in the case of the test-tube baby, let's say, the baby, in effect, *is* the test tube.

UB: Yes, that's right. The baby is the experiment. This has changed the role of science in our society. You have to do it; otherwise you can't find out about it. I sometimes feel a certain pity for the scientists because they are in a situation in which they have to ask for money and claim that everything is certain and that everything is being done in the proper way. In fact, they don't know what the consequences of their action will be, and they cannot know before they do it. Thus they are in a very complicated situation. Sociology is in a much more comfortable situation because we are not changing society in order to test our theories. But scientists have to change society, and they have to change the environment in order to test their theories. This very complicated situation can create severe ethical problems and questions.

DC: Many of these ethical questions are not currently being addressed, and the reason, Beck thinks, is that we lack the necessary institutions and the necessary criteria for assessing new technologies in advance of their use.

UB: It is a fate. We are constructing a fate. Technology is produced and we have to use it. New technologies come up. We have a discussion about whether or not we will use them. We consider the ethical implications. And none of this matters at all. The technology comes anyway. If it is good for the market, if it sells and so on, it's coming. This is a kind of fatalism; it's not modernity. We are not able to decide what technology we want and what technology we don't want. We have to take it. There is no democratic decision-making in relation to technology.

DC: Modern societies, unlike all previous societies, were supposed to be rational, but instead they have embraced technology as blind fate. In this sense, Beck argues, modernity has turned out to be radically unmodern. This is the basis of his call for a new self-critical modernity that has the institutional capacity to look before it leaps. Creating this second modernity will require us to overcome many of the mental habits of the first modernity. One of these is the habit of cutting the world up into simple opposing categories or dualisms.

UB: So far we have thought about modernity and organized modernity in specific dualisms — society and nature, for example, or we and the other, and so on. One of the amazing developments we can show in all kinds of different fields is that, as a result of the radical effect of modernity, those dualisms don't work anymore. We have a hard time making clear distinctions between nature and society. The ecological crisis is an example. When we talk about climate change, we're talking about nature, on the one hand — it's a question of rain and ice and all sorts of seemingly natural things — but at the same time, we know that what we are facing is a consequence of modernization. In all fields, under the radicalizing influence of modernity, basic distinctions are being undermined, and we have a new mixture.

DC: This new mixture is exemplified by the phenomenon of climate change. Climate change is an entirely natural phenomenon. The carbon dioxide I produce is no less natural than the emissions produced by a forest fire or a volcano.

Yet it's entirely social, since I produce it in pursuit of socially determined purposes. Climate change is also a phenomenon that is both real and constructed, to take another outmoded dichotomy. Anyone can recognize a melting glacier or a run of unusually hot summers, but they can only be recognized as climate change through the lens of an elaborate and fallible scientific construction.

UB: The sciences make risks visible, they make invisible risks visible. Climate change is an example. We cannot actually see climate change. We cannot feel it. We cannot experience it. Even if, as we find out now, the weather is changing dramatically — maybe in a few years people in New York will be wearing swimming clothes at Christmas — even so, there's a difference between weather and climate. Climate is a very general concept which has a completely different point of view. It includes all kinds of statistics and models and measurements that take in thousands of years. We don't actually see climate change. We think we see it because we have the interpretive model of climate change in our minds, and then we see the differences. However, this all has to be constructed.

DC: Constructed knowledge is, inevitably, fallible. The construction may be faulty or the object of knowledge may be too complex to comprehend. Either way, risk society faces an imponderable and uncertain prospect. Which risks should we attend to? Which sources of knowledge about these risks should we trust? This uncertainty is what makes risk society so intriguing to Beck as a sociologist.

UB: We are confronted with the anticipation of catastrophe, not the catastrophe itself, but the anticipation of catastrophe. In order to avoid the catastrophe, we have to take the anticipation seriously, even if we don't know whether the catastrophe is really going to happen. There are two kinds of mistakes one can make. The first mistake is taking the anticipation seriously, maybe getting hysterical about it and over-dramatizing it, and nothing happens. But then it could happen the other way around, too, couldn't it? We don't take it seriously. We don't anticipate. And then the catastrophe happens. This makes risk so interesting, sociologically, so interesting. There's a lot of non-knowledge involved. We have only the anticipation of a catastrophe. Nobody really knows what all the relevant causes and relationships are, and we have to react in a situation where we don't know what's really going on in order to avoid the potential of a catastrophe. If we don't do it, then maybe it's going to happen. This is true of climate change. In order to avoid it, we have to act now. However, if we do act now, it's possible that nobody will ever know if climate change was the right diagnosis.

DC: How can one live a decent, local, embodied, primarily face-to-face life in the midst of all these virtualities?

UB: Well, the sociologist is first of all fascinated by the phenomenon and doesn't always have the solution for it. You know? We are the ones who analyze these situations. First of all, I think that risk and the anticipation of catastrophe is an amazing phenomenon which is very close to irony and to drama. People have to act and institutions have to change, but nobody really knows what is going on. It's a fantastic, realistic/unrealistic, permanent state of I-don't-know-what — maybe it's a play, a piece of theatre played in reality, and we're all playing this role or that role in our political life and in our daily life.

Now, if you ask me as a person, not as a sociologist, but as a person — as an individual — this is a point where I really have to confess that I think that a worldwide risk society, with these global risks, is tragic for everyday life. It has tragic implications for everyday life. On the one hand, even if we try to reorganize our lives by riding a bicycle rather than driving a car, by not taking vacations overseas but just going around the corner and having a nice time, it's not necessarily going to change things. Of course, we can say to ourselves, if everybody does it, maybe there's going to be a change. I am not saying it doesn't have an implication, but it is actually not a basic change. We don't have the elements. We don't have the institutions. We don't have the answers as individuals. Many of the dangers are not even part of our own experience. We cannot see them. We cannot smell them. We cannot taste them. We cannot feel them. We need expert interpretation, expert rationality. So it's quite a dilemma. On the one hand, we try to get away from expert knowledge, but, on the other hand, it's on the basis of expert knowledge that we know we need to get away from it. We have to make decisions, but there are no real answers to the dilemmas. As a person, I would say there's still one basic answer. Don't take things too seriously. Be ironical about all the different issues. Otherwise you lose your autonomy. Otherwise you lose your own judgment. Otherwise you lose your own way of life.

DC: Daily life in risk society thus has both a tragic and a paradoxical element — tragic because of our individual powerlessness and paradoxical because of our dependence on experts to get us out of the problems experts got us into in the first place. But risk society may also be a forcing ground for a new modernity, a second modernity, with the institutional capacity to actually deal with the mess that modernization has made.

UB: We are living under conditions of organized irresponsibility. If you ask, who is responsible for climate change?, you'll have a hard time answering. The scientists are not, the economy is not, the politicians are not, the intellectuals, of course, are not. Nobody is actually responsible. This is the current institutional answer to the question of responsibility. I think, therefore, that in order to get out of this mess, we have to readdress the question of responsibility, and we have to re-address the question of global justice as well. We know, for example, that those countries and regions which are most affected by climate change have had the least impact on the climate, and they have a certain right to develop and start industries in order to become part of modernity. In order to solve these problems, we have to include such considerations and ask ourselves what is going to be a just way of redistributing the economic benefits and ecological costs of development. We need a frame of reference to talk about responsibility and global justice. Only if we do this will we find solutions on the national level. This is the new political utopia, you might say, which is implied by our search for an answer to the climate change problem.

DC: You call this a new utopia?

UB: I call it a new utopia. It's not the old utopia, because it's not a political project produced by a political decision. It is something which faces us as necessity if we are to solve the problems of future generations and the reproduction of life on earth. It is an enforced cosmopolitanism. It is an enforced utopia which we need in order to solve our national problems. Risk enforces a public discussion, maybe, beyond local and national borders. It creates, let's say, a training camp or a learning process for the creation of new normative principles which I would call cosmopolitan principles. Now, we may have to redefine the word cosmopolitan. For me, it means not only, as we say in German, *der Weltbürger*, the world citizen, which is a concept of the Enlightenment, a concept of Kant's. It's something different. It is something which is being enforced, which is not necessarily wanted but is happening as a result of our being engaged in global risk scenarios. It's not something which people vote upon, not something which they choose, not something which they reflect upon in a very sophisticated way, but something which happens to us. Suddenly we are living in a situation where we cannot exclude the other any more. Even if people don't want this to happen, even if people are looking for a world which is less connected, which is less cosmopolitan, they are still reacting to this basic situation. Actually, they are affirming this situation to some extent: it is very much part of the nervous breakdown of the world in which we cannot exclude the other anymore that people should try to find more sophisticated ways of building walls and borders. But they don't work anymore under conditions of global risk.

From Critique
to Composition

Why is it that, when you describe science and the whole interesting range of things which are required to produce a sturdy fact, people believe the description is debunking? What's their idea of society when mere description of science becomes a threat?

— Bruno Latour

Ulrich Beck says that modernity requires modernization; he proposes a "second modernity," with institutions that are cosmopolitan, responsible, and self-aware. French thinker Bruno Latour views the same predicament in very different terms. He argues that we have never been modern. In fact, he gave that exact title, *We Have Never Been Modern*, to one of his best known books. His position is both like and unlike Ulrich Beck's. Like Beck, he claims that modernity has been based on an artificial separation of nature and society and that this separation has now been revealed as a myth by the complete entanglement of nature and society in phenomena like climate change. We can now see, Latour says, that there never really was a pure "out there" called nature or environment, nor was there ever a purely social reality standing apart from this mythical nature. Nature has always been part of society, and society of nature. But this insight pushes him to a conclusion very

different from Beck's. Beck wants a grown-up modernity. Latour wants
to abandon the whole idea.

Bruno Latour is a professor at the Paris Institute of Political Studies
— Sciences Po, as it's called — which has long trained the French
political elite. I spoke with him there, and he told me a little about how
he came to view modernity as an outworn political myth. Beginning
in the 1970s, he and a few others, working independently, began to
do a new kind of research: close study of how scientists actually go
about establishing matters of fact. Up to that time, the philosophers who
had reflected on the nature of science had not been interested in this
kind of description; philosophy had treated science as pure knowledge,
uncontaminated by the social interests and opinions characteristic
of politics. This partition, Latour says, with science on the one side,
politics on the other, left no room for realistic description.

BRUNO LATOUR

Philosophy of science never was descriptive. I mean, its aim was not
scientific, in a way. It was always political, in the sense that it had to
establish a certain definition of what science was for pedagogical and
political and polemical reasons. It said, there is one question which is
about science and another question about politics. There is one way of
solving the question for politics and another one for science. That's what
I call political, because it's what makes the divide between the two.

Of course, this is a very old idea that goes all the way back to the
Gorgias of Plato. It has a very, very long history, and I've tried to
write about some of it. At the same time, the whole operation is not
to describe the practice of science, but to make a distinction by saying,
okay, you guys in politics, you are talking only about opinions and
about subjectivity and about ideology, whereas in science there is only
demonstration and, basically, rationality. This position, which makes
the separation between the two, is entirely political. It organizes a
political discussion by dividing the field, as I showed in *The Politics
of Nature*, into two entirely different ranges, so to speak. This very
divide between science and politics is a political invention, originally a
Greek invention, but then constantly reinstated by many, many different
groups. All the way from Plato to [Steven] Weinberg, it's the same

argument being rehashed. On one side there is politics, and on the other side is demonstration and rationality and so on. This view entirely blocked, or interrupted, the task of describing science. There was no space for description, there was no interest in laboratories or computers or the social networks of scientists, no interest in how scientific facts emerge — all of these things which the group I belonged to found so fascinating. In a way, you could say that our project was not to politicize the sciences, as some said, but more to free the sciences from politics.

DAVID CAYLEY

Latour and his colleagues set out to describe the real world of science: its social relations, its economy, its technical apparatus, its lines of communication, and so on. This freed science from politics, in his terms, because it replaced the political myth of science as pure reason with a picture of a discipline as vivid, messy, and interesting as any other realm of human endeavour. But where had this myth come from? Latour traces it all the way back to Plato, but he thinks that the variant we have lived with goes back to the seventeenth century, when a society, scarred by the endless warfare of opinion, tried to establish a new source of certainty.

BL: It was the time of the idea that finally we were detached from all the shackles of the past, so to speak. We were entering the domain of reason and mastery and control and all sorts of nice things. We were right in believing in it, but now that domain has disappeared because we have realized that we are entangled in a way which we could not even have imagined before. We have shifted from a history in which we were being emancipated more and more every day to a history in which, by being emancipated more and more every day, we are now entangled, in even more important ways, to the point of having the climate on our shoulders. I mean, if you had told that to Descartes or Kant or Hegel, though Hegel might have been amused, they would have thought you were speaking only metaphorically. They would have said, well, you are deluded. Humans are not strong enough to have any effect on the climate. This is what people believed in the past, they would have said, but now we are finally free from these old myths. Now, though, it's not a myth. Now you have a Security Council meeting to decide what sort of threat global warming presents. The future is no longer one of emancipation. It's a future of attachment or care or whatever expression you want to use.

DC: Modernism, for Latour, was the attempt to rise above nature on the wings of reason, but, like Icarus, the higher the modernized world flew, the more quickly its wings melted. Today, complex entanglements like climate change make it obvious that the future belongs to an ethic of care, not of domination. Modernity was based on the idea that all agency belongs to human beings — the world as an object of scientific knowledge has no voice, and it has no influence on how it is known, since we know it, as we say, objectively. But this is all nonsense, Latour says; we know perfectly well that things do influence the ways in which we can know them. One of the examples he develops in a book wonderfully titled *The Pasteurization of France* is the relationship between Louis Pasteur and the bacteria, the microbes, that he made known.

BL: The discovery by Pasteur of microbes was a very important event in the history of civilization — it's a large reason why there are now nearly seven billion human beings — but it was also a very important event in the history of microbes. Something happened to them and to our relation with them, thanks to Pasteur and because of his laboratory, something that had never occurred to them before, so to speak. We are still in this history because we are still fighting with microbes, which are transforming themselves quite fast; a very nice illustration is the resistance they are developing to antibiotics. We are still are in an ongoing history, which is a mixed history of microbes and of scientists discovering microbes.

DC: This mixed history of people and microbes is a crucial idea for Latour. What he calls the modern constitution was based on a series of purifications — separations — of society from nature, humans from non-humans, and, later on, subjects from objects. At the same time, it is plain, to stay with Latour's example, that microbes no longer fit any of these categories. In the age of superbugs, no one will think them mere passive objects. It makes equally little sense to call them subjects. They belong to nature but no less to society, since it is society that creates the terms and conditions according to which they live. The modern constitution, therefore, is a fiction, Latour says, and many scientists recognize this.

BL: In practice, all scientists know perfectly well that they are dealing with very complex and interesting mixtures. It's a very complex relation that you cannot cut by saying, okay, choose: either the facts speak for themselves, or you speak for the facts. If you make that choice, you become really silly, because then you deny the role of laboratory practices in shaping scientific objects. The agency of technology and the agency of things is, by now, completely obvious. Again, this is a point that I made thirty years ago with some trepidation, because I knew I was passing across a divide which one was forbidden to pass by the philosophy of science. Today it's completely obvious that there is no such divide.

DC: The divide that Latour crossed separated science from the facts that science establishes. Under the modern constitution, an established fact was a free-standing reality. Once discovered, scientific laws were assigned to nature or to God, and all marks of their human origins were erased. Latour argued that facts remain attached to the practices and assumptions that make them facts in the first place. A fact about climate, let's say, depends on the apparatus that made the observation. It depends on the network of definitions that stabilize the meanings of basic terms in meteorology. It depends on the good faith of the researcher who attests to its veracity. It depends, in short, on what Latour calls science practices. These practices do not invalidate facts; rather, they make the facts what they are. But this was not, at first, appreciated.

BL: When I did this work on science practices, no one understood it. It was taken as a debunking of science. I was interested in that reaction because I never thought that science had to be debunked. I thought it had to be studied and described, but debunking never interested me. Yet my work was taken by people as debunking. I became interested in that argument. Why is it that, when you describe science and the whole interesting range of things which are required to produce a sturdy fact, people believe the description is debunking? What's their idea of society when mere description of science becomes a threat? I realized that when people were threatened by a description of science, it was because they had already entirely separated social explanation for things they didn't believe in from science and rationality, which explained the things they did believe in. The reason my colleagues and I seemed to be so threatening was that people believed that when you give a social explanation of something, you debunk it. A social explanation of religion implies that there is no religion. Social explanation of art? There is no art. Social explanation of politics? There is no politics. Social explanation of law? There is no law. The only exception was social explanation of science. Ah, there you had to backtrack, back-pedal entirely, and say, no, social explanation of science is not possible because . . . science is objective. So it has no social explanation. I was very interested in digging out this completely perverse definition of the social sciences, which is really at the heart of the social sciences. Basically, when they study something, it disappears. They have the King Midas touch, but in reverse. Instead of transforming everything into gold, they transform it into dust. It's so odd, so extraordinary, the belief that the phenomenon they describe, the object they describe, does not really exist.

DC: Latour gave his first book, *Laboratory Life*, the subtitle *The Social Construction of Scientific Facts*. He soon realized that he had unwittingly said something quite different than he intended — that to say something was socially constructed was, in effect, to make it disappear. He had hoped to enrich the sciences by a thicker, more vivid description of what they actually do. He concluded that the whole modern world picture in which a purely social society stands apart from a purely natural nature was a political myth devised by those who could claim to speak for these carefully segregated realms, the sciences for nature, the political delegate for society. This myth, he says, is now clearly finished, undone by an ecological crisis in which human and non-human agencies are clearly blended.

BL: We are engaged in a huge redefinition of what agency is: our own agency vis-à-vis climate and climate's agency in relation to us, for example. All these questions that were supposedly solved are now open again. This reopening of the question of agency in terms of climate and in terms of the ecological crisis more generally is one of the things that has turned a position — my position — which seemed strange earlier on into common sense. It's completely common sense that humans are not the only ones who have agency in the world. I mean, it's only if you have read too much Immanuel Kant that you can now believe the opposite.

DC: The task of this new age, in Latour's view, is composition, not critique. Critique, for him, belongs to the revolutionary ethos of the modern period. Critique is breaking things down, finding the truth behind the appearances, getting to the bottom of things. But it is time now to put things back together.

BL: Critique is past. We are in such a situation of ruins, intellectual ruins, that the question has now become one of composition. See, the big difference from the time of revolution, the time when we believed that revolution was a way to do things, is that things now have to be done in detail. The whole still matters, but it has to be considered in detail. That's very striking. If you take an ecological argument — whether it's about rivers or localization or delocalization — the small details are what matter. It's not a question of a revolutionary shift. Composition means that you have to take up all the tasks of assembling disjointed parts, so to speak, from the bottom up. It's true in art, it's true in politics, it's true in science. It's true in many different ways.

Through Gaian Eyes

JAMES LOVELOCK

Among the Christmas presents I received at four years old was a box of bits of electrical gear that my father put together. The interesting thing was, I remember afterwards going round the house and asking everybody, why do you need two wires to send electricity along? You can send water along a single pipe. Why should you need two? Of course, nobody could answer me. Not even the postman, who was very knowledgeable otherwise. It was then I realized that if I wanted to know the answers to difficult questions like that, I'd have to find out myself. And that set me on the right course in life, I think.

— James Lovelock

Forty years ago, British scientist James Lovelock put forward the first elements of what he would come to call the Gaia theory. Named for the ancient Greek goddess of the earth, it held that the earth as a whole functions as a self-regulating system. At first, many biologists scoffed. Today, Lovelock's ideas are more widely accepted, even in circles where he was initially scorned. But even as he has been winning scientific honours, he has been growing more pessimistic about the prospects for contemporary civilization, and in 2006 he published *The Revenge of*

Gaia, which predicts a dramatically hotter and less habitable earth by the end of the present century.

Science, in our time, has grown so specialized and so professionalized that it's almost shocking to think that the term "scientist" has designated a profession for little more than a century. Michael Faraday, the discoverer of electromagnetism, whose career spanned the first half of the nineteenth century, still called himself a natural philosopher. It wasn't until the end of that century that the word "scientist" was regularly used to name a possible career. In France, to take just one example, there were fewer than three hundred students enrolled in science faculties in 1876. Forty years later, at the beginning of World War I, there were more than seven thousand. With this professionalization went specialization and sub-specialization, to the point today that science is comprised of hundreds, if not thousands, of tiny non-communicating realms, and the astrophysicist may know nothing more about molecular biology than what appears in the newspapers.

Jim Lovelock is a bit of a throwback in this hyper-specialized world. He's remained a generalist whose work often defies disciplinary boundaries. He makes his own scientific apparatus, and for nearly fifty years, he's pursued an independent career in a scientific milieu now completely dominated by institutions. I've interviewed Jim Lovelock a couple of times for *Ideas* since he first put forward his Gaia hypothesis, and, in the fall of 2006, I had the pleasure of calling on him, for the first time, at his home, a restored mill in the beautiful countryside of Devon in southwestern England. He began by telling me about his upbringing in the south London district of Brixton, where he was born in 1919 and where his parents kept a small art shop and gallery. I asked him how his interest in science developed.

JAMES LOVELOCK

Brixton had a wonderful library. In those days you didn't buy books, you got them from the library. I used to go down as a small child, and they allowed me to borrow books in spite of my being quite young. I soon started reading science fiction, things like Jules Verne and H.G. Wells. Wells was an absolute revelation. I suppose he was the first really modern science fiction writer. But after a while, I tired of it and went for the hard stuff. In the basement there were textbooks of all the various sciences, and I started taking them home. Jeans's *Astronomy and Cosmogony* was one book I took, and then there was Wade's *Organic Chemistry*. I used to read those avidly as a kind of devotional reading.

I don't suppose that at eight years old I understood more than a fraction of it, but children of that age group, pre-puberty, take in information almost as machine language, and it stays for life. You never lose it. It's all there. I think I built up an enormous repertoire of scientific information right across the board, from astronomy to zoology inclusively. When I went to grammar school at age ten, I found the science taught unutterably dull, almost pointless. So I just went on reading by myself.

DAVID CAYLEY

After he completed his schooling, Jim Lovelock went to work as a lab assistant for a firm that did scientific consulting for the photography industry. He would be dispatched, for example, to find out why a certain batch of the gelatin used to coat film lacked the required sensitivity to light. This apprenticeship in practical laboratory technique proved invaluable.

JL: One thing I learned at this consultant's was that in doing an analysis, which we did quite a lot of, you've got to get the answer right. There must be no fudging and no excuses. This is the opposite of what's taught in universities, where the student is told that, in doing science, it doesn't really matter if you don't get the right answer, as long as you understand the method. That'll be sufficient for the exam. The purpose, the objective, is to pass the exam, not to get a correct answer. This is fatal for scientists. It leads to all manner of trouble. I was very fortunate to have what many would call a deprived start, instead of going straight to university.

DC: Lovelock did eventually go to university, training in chemistry at the University of Manchester. A curious sidelight of his career there was that he quickly faced a false accusation of cheating because one of his professors found his lab results suspiciously precise. Lovelock then got a job with the National Institute for Medical Research in London, and he worked there as a research scientist for twenty years.

JL: I went to the National Institute in 1941, during World War II, and I left in 1961. The reason I left was that it was such a wonderful institute to work at. I had tenure, I had an exceedingly good income, it was almost ideal. I could work on any subject I liked, as long as it had a connection with medical research. But I could see tramlines of security going down to retirement and the grave, and I couldn't stand it. I was fretting and trying to think of what to do and how to explain to my bosses that I wanted to leave on such trivial grounds when out of the blue came a letter from NASA, which was then only three years old. It was from their director of space-flight operations, asking if I would be an experimenter on their lunar and planetary missions.

DC: Lovelock went to NASA, the National Aeronautics and Space Administration, and from then on worked as an independent scientist. The reason NASA wanted him was that in 1957, while still at the Medical Research Institute, he had invented something called the electron capture detector, a device capable, for the first time, of measuring chemicals in the trace amounts, the parts per million, billion, or even trillion that we are familiar with today. A thousand times more sensitive than any existing device, the electron capture detector became, as Lovelock says in his autobiography, the midwife to the environmental movement that emerged in the 1960s. It was the electron capture detector that measured the accumulations of DDT that Rachel Carson brought to public attention in her 1962 book *Silent Spring*, and it was the same device that detected the buildup of ozone-threatening chlorofluorocarbons, or CFCs, a decade later. In that case, it was Lovelock himself who took the measurements from aboard a British research vessel in the Antarctic.

But back in 1961, NASA's Jet Propulsion Laboratory, known as JPL, wanted Lovelock to help design the device that would analyze the surfaces of the moon and Mars. He soon grew interested in a larger question: how does one test for life on Mars in the first place? In that question lay the seeds of the Gaia hypothesis.

JL: Looking for life on another planet always seemed to me to be a rather difficult thing to do, to say the least of it. You've got to go there and start scrabbling around in the soil. And how do you know what the life is? It mightn't be the same as here. You mightn't be able to recognize it. I thought about this when I was at JPL, and I said, look, isn't this the wrong way to go about the whole thing? If there is life on Mars, it'll be bound to use the atmosphere as a mobile medium to supply itself with raw materials, and that will also be the only place it can dump its wastes. Now, doing that will change the atmosphere chemically in a way that should make it recognizably different from the equilibrium atmosphere you'd expect of a dead planet. This led to a great battle in JPL. All the physicists and chemists saw this as a great idea and supported it. The biologists just hated it. They didn't think that was the right way to look for life.

The battle made me think about the earth's atmosphere. Does its composition prove there's life on the planet? Yes, it does so, in spades. If you look at our atmosphere, you've got oxygen and hydrocarbons mixed, and that's unstable, chemically. I mean, it could never have happened by accident. Indeed, it's so unstable that, if you just left it alone, if you isolated the earth and took all life away, the methane in the atmosphere would halve in about twelve years. What that meant was that there must be something on the planetary surface that is making both methane and oxygen at a rate that sustains the observable concentrations, and these are huge quantities. Obviously, what was making them was life. Indeed, there was evidence already that organisms made methane. We know they make oxygen. So here was a very good life detector, but it also meant to me that something must be regulating the amount made; it couldn't by accident have stayed constant for millions of years, as it appeared to have done. This is what made me think of Gaia.

DC: So the innovation, in a way, was not the discovery of the composition of the atmosphere, which was known, but the question, how did it get that way?

JL: Exactly. You're the first journalist to have recognized that, and I'm grateful to you for having said so.

DC: What shifts is the point of view. You were the first to look at the earth, in effect, from Mars.

JL: Well, I had been asked to look for life on Mars, and I had suggested that examining the atmosphere was a way of looking for it. My mind was primed and ready to look at the earth's atmosphere that way, whereas had NASA not existed and raised the question, how do you look for life on another planet?, I would never have asked it about the earth. Nor would anyone else, I think.

DC: Lovelock had suggested to NASA that the way to test for life on Mars would be to ascertain the composition of its atmosphere. In September of 1965, a French observatory produced the relevant data. Mars's atmosphere turned out to be completely inert, the exact opposite of earth's lively and volatile envelope. "At that moment," Lovelock recalled in his autobiography *Homage to Gaia*, "my mind filled with wonderings about the nature of the earth." Astronomer Carl Sagan was with Lovelock at the time, and Lovelock blurted out to Sagan his intuition that living things at the surface of the earth must regulate the planet's atmosphere, else how could such an unstable mixture be kept constant? Sagan was skeptical, but he did impart a second crucial piece of information. The sun, he told Lovelock, is now thirty per cent hotter than when life on earth began, and yet the temperature of the earth has remained relatively constant during that time, varying enough to produce periodic glaciations but nothing like what a thirty per cent change in solar output would predict. To Lovelock, this was further confirmation, suggesting that the earth must regulate its own climate, as well as its atmospheric chemistry. "Suddenly," he says, "the image of the earth as a living organism emerged in my mind." So grand an idea required a grand name, and that came from novelist William Golding, the author of *Lord of the Flies* and winner of the Nobel Prize for literature. He and Lovelock were neighbours in a village in Somerset during the 1960s.

JL: He was a lovely man. We often used to walk together, chattering. He was originally trained as a physicist, so he knew about science. He was very interested in what I was doing at JPL and wanted to be kept up to steam on it. I was telling him about this idea, and he said, well, if you're going to have a big idea like that, you better give it a proper name. I said, okay, what shall I call it? He said, I'd call it Gaia. I thought he meant g-y-r-e. We walked on for nearly twenty minutes, talking at cross purposes, until he finally said, what do you think I meant? I explained what I had thought, and he said, no, no, no, no, I meant the Greek goddess of the earth, Gaia, you know, the root of geology and geography and so on. Well, when you get given a name by a wordsmith of quality like Bill Golding, you don't turn it down. But boy, has it given me trouble.

DC: The name of the goddess was a blessing and a curse at the same time. It attracted wide notice and wide discussion among the public, it delighted the hippies, feminists, and environmentalists who wanted to revive Mother Earth, but it also made it easy for some of his fellow scientists to dismiss the theory as unscientific. Biologists were particularly scathing.

JL: They really trashed it in a big way. Some of the senior members of the biological community, like John Maynard Smith and Bill Hamilton, told me afterwards that they'd never even read the papers or the books. They had just picked it up by hearsay and regarded it as in the same category as creationism — another attack on Darwin — and they were out to smash it at all costs.

DC: The hostility of biologists proved long-lived. In an article in the journal *Science* in 1991, for example, the late Stephen Jay Gould was quoted to the effect that Gaia is a pretty metaphor and not much more. Despite this inhospitable climate, however, Lovelock continued to elaborate his theory and to test its predictions.

JL: If it's a good theory, it makes good predictions and suggests experiments to do. Gaia was particularly fruitful, and still is. It suggests experiments you can do to prove it. For example, if I was right, then the transfer of elements from the sea to the land, which is very important, must be regulated. What elements would be useful to look at? Well, two elements in particular: sulphur and iodine. They're both quite scarce on the land, too scarce for plants, and iodine is too scarce for animals, but they are transferred from the sea because the sea has them in rich abundance. So is biology involved in that process?

DC: Do creatures in the sea make sulphur and iodine in the quantities required by creatures on the land? That was Lovelock's question, and he was able to answer it during a voyage he took on a British research vessel called the *Shackleton* in 1973. His primary purpose was to measure the atmospheric concentration of the chlorofluorocarbons that were suspected of damaging the earth's ozone layer, but he also took readings for the volatile compounds of sulphur and iodine that are emitted by plankton and seaweed in the oceans. He found them in an abundance that suggested that this was the source from which the land was being supplied.

JL: The prediction was confirmed, and somebody took my results — a scientist at East Anglia called Peter Liss — and calculated the rate of transfer from my measurements of the amount of these two gasses in the air and in the sea. Sure enough, it was exactly right to fulfill the sulphur cycle and the iodine cycle, and the theory was confirmed numerically.

DC: Dimethyl sulphide and methyl iodide are made in the sea, then transferred to the land in just the quantities required by terrestrial organisms. But perhaps this is just a happy accident, a bit of luck for the iodine- and sulphur-deficient life on the land. Something like that was the prevailing view of a lot of the processes·that Lovelock believes are part of Gaia's self-regulation. They just happen. Geologists, for example, had held that the burial of atmospheric carbon dioxide as limestone was a purely geological process, but Lovelock was able to show that organisms accelerate this process to exactly the point that is necessary to maintain a stable temperature.

JL: The geological wisdom before Gaia was quite simple. As it gets warmer, more water evaporates and more rain falls on the land, and the rain contains carbon dioxide dissolved in it. The carbonic acid in the rain attacks silicate rocks and dissolves them to form calcium bicarbonate and salicylic acid. Those go down the rivers into the sea, and the calcium carbonate gets deposited as limestone. That's what pulls the carbon dioxide out of the air. It's a pump. As things get warmer, more water evaporates and more rain falls, so the rate of pumping increases, which tends to pull the CO_2 down. That is a self-regulating system, but it's purely geological.

It's a plausible theory, but when you put the numbers in, you find out that the earth's temperature should be somewhere in the region of sixty degrees Celsius. The theory is obviously wrong; that temperature is far above what it should be. However, all you have to do is to say, oh, it's not that simple. Organisms growing on the rocks and trees growing in the soil enormously increase the rate of attack of carbon dioxide on the silicate rocks. They break up the rock. If you measure the CO_2 in the soil, it's thirty times higher than in the air. Thus the biology is also working to pump it down.

DC: Geologists had recognized that rock weathering removes carbon dioxide from the air, but they had overlooked the extent to which biological activity in soils and on rock regulate this process. If just the action of rain on rock were considered, the earth would be unlivably hot, but when the contribution of living things is factored in, a sensitive temperature stabilization system emerges. This was one of the first Gaian mechanisms to be worked out. Others followed.

JL: The really exciting one — one of those rare moments that make science so worthwhile as a life — came in Seattle in . . . I suppose it would be about 1985. I had been invited to spend a month at the university there on some sort of special fellowship deal, and while I was there, I gave a lecture on Gaia. A chap who was an atmospheric chemist, Robert Charlson, came up to me afterwards, and we got taking. He said, I was very interested in what you were saying because we've got a very big problem. He said, we know how clouds form. They need tiny droplets of some nuclei in the air on which the water can condense and grow. I said, well, what's the problem? What are the nuclei? He said, well, they used to think it was bits of sea salt that had been, you know, sprayed up from waves breaking, but we've been flying planes all over the Pacific, far away from the land, and we always find the same thing: it's droplets of sulphuric acid or ammonium sulphate, never sea salt, or very rarely sea salt. Where's it coming from? It can't come from pollution because pollution falls out fairly quickly, and we are finding this phenomenon right out in the middle of the Pacific, far from any land source. It's got to be local. It can't come from volcanoes, because there haven't been any when we were flying. So where does it come from? Immediately, I thought of the dimethyl sulfide. I said, I think I know. It's the gas that comes out of the sea, the biological gas. It is oxidizing up there and forming sulphuric acid. Sure enough, that was proven up to the hilt, because at the same time in these little droplets there's an acid called methanesulfonic acid, which can come only from the oxidation of dimethyl sulfide.

DC: And who or what is making the dimethyl sulfide?

JL: The organisms in the ocean. And the clouds, of course, regulate the climate. Without the clouds, the earth would be eight degrees warmer.

DC: Because they reflect light away . . .

JL: Yes, and there's a Gaian mechanism that's really solid. It was so solid that within a year the World Meteorological Organization gave us the Norbert Gerbier Prize for the most significant bit of meteorological research of the year. But the biologists kept right on trying to trash it and prove that cloud formation couldn't possibly have anything to do with climate regulation — a quite extraordinary affair.

By the mid-1990s, I got fed up with this endless battle with biology, and I started thinking, what have you done wrong? Most battles take two sides to continue, so what have you done that's brought it about? Then it suddenly dawned on me. My cardinal error had been that I'd been talking all the time to the middle management of biology, never to the leaders. I immediately went to see the leading biologists in this country, who were Robert May, a man called Maynard Smith, John

Maynard Smith, and William Hamilton. They were the key figures in the neo-Darwinist movement. As I think I mentioned earlier, they said to me, well, to start with, we think your theory's nonsense. Then I said, why? Have you read the papers? They admitted that they hadn't. I've got a letter from Bill Hamilton apologizing that he hadn't had time to read any of the papers. He had gone by what his graduate students told him about it. And that's classic. That's the way these people work.

When I was able to explain to them, they swung right round and said, oh, now we understand what you're talking about. Yes, we agree, the earth probably does self-regulate, but we cannot, for the life of us, see how this can happen by natural selection.

DC: The objections of biologists turned on the perceived threat to Darwinian orthodoxy. Organisms are supposed to do what's good for them and benefits to others are incidental, but what immediate practical advantage could phytoplankton in the sea possibly derive from fostering cloud condensation and thus helping to keep the earth cool by reflecting away light? How could natural selection ever have produced such an arrangement?

This was the crux. Lovelock had shown persuasive evidence of planetary self-regulation, but it was clearly impossible to explain on narrowly Darwinian grounds. His answer was, expand the explanatory framework to take account of the new phenomena.

JL: I think the neatest way of looking at it, which does sound a bit hubristic, is to say that Newtonian physics was pretty well perfect, right up until people started looking out into the cosmos, or looking down into tiny particles. The same is true of Darwinian biology. It worked perfectly. It's only when you push it to an extreme, to the scale of the whole planet, that you come up against its limitations. Then you need an additional theory, and that's what Gaia is all about — expanding Darwinism to include the regulation of the earth. You have to look upon the earth itself as a living organism, but not of the same kind as the ones that participate in natural selection. Gaia stands to Darwinism as relativity stands to Newtonian physics. It doesn't displace or disagree with Newtonian physics. It just extends it.

DC: Biologists rejected the Gaia theory because it seemed to them to imply that the lichens that accelerate rock weathering or the plankton that furnish nuclei for clouds had some sort of larger purpose in view, whereas modern science is virtually defined by its having excluded any ultimate purpose from its purview. That was the old Aristotelian language in which each thing had its proper end, its good, its purpose. Modern science was built on mechanical explanation. Evolution, for example, has a mechanism, natural selection, but no goal, no ultimate end. But now, here was Jim Lovelock claiming that the entire biosphere is a massive collaboration with a definite end, the maintenance of an agreeable environment for life. How could this view be reconciled with the basic postulates of modern science? The key lies in understanding the difference between mechanical systems, clockworks of various kinds, and self-regulating systems.

JL: Engineers who talk about self-regulating systems, like autopilots or refrigerators, things like that — they happily talk about goals: the goal of the system is such-and-such. Now those, of course, are fighting words to biologists. They say, oh, an engineered system is different because you set the goal of your refrigerator. It doesn't have to find the goal itself. Somebody sets it. Then take a step up and ask, what about you? You regulate your temperature well, but which of you — to use the Biblical phrase — by taking thought can alter your temperature by one degree? Of course, the answer there is that the goal is set by genetic memory and evolution. You have evolved as an animal that operates best at that temperature, and that's that. The goal has been set by your historical evolution. When you come to Gaia, the goal of Gaia for self-regulation is set by the properties of the universe. You see, our life is all carbon-based. The carbon compounds that make up living cells, particularly their membranes, are quite fussy about the range of temperatures and conditions at which they can exist. It's those that have set the goal of the self-regulating system Gaia. It had to evolve that way because any organism that broke the rules wouldn't leave any progeny. Natural selection, obviously, picks the ones that do it right. You don't really have to look any further. It's the properties of the universe that set the goal. But there is a goal.

DC: Self-regulation is a hard idea to grasp because a self-regulating system is, in effect, its own cause. Once a thermostat is set, the temperature regulates the furnace and the furnace regulates the temperature in an endless circle. However, Lovelock thinks that the mind prefers linear explanations in which distinct causes clearly precede their effects.

JL: There are two classes of phenomena we know about which are mentally baffling. Self-regulation is one of them. Perhaps our greatest physicist ever was James Clerk Maxwell. He went to an exhibition — in 1850, I think it was — at the Royal Society, and there he saw [James] Watt's steam-engine governor. I don't know whether you know the thing. Steam engines were all the craze then. That was high-tech in the 1850s. This governor was a device which kept the steam engine's speed constant. It consisted of a couple of iron balls that spun out from the engine. The faster the engine went, the further the balls spun out. This lifted a lever that put negative feedback on the steam supply to the engine, so it automatically set itself at a constant speed. It's a very simple device. I remember seeing it as a child, when I went to the science museum, and instantly knowing how it worked. Three days after he'd seen it, James Clerk Maxwell reported at a Royal Society meeting — I think it was a council meeting — that he'd had three sleepless nights trying to work out how that governor worked mathematically. It defeated him. Later, he published papers in which he claimed he had represented the function of the governor mathematically, but it was really done by a mathematical fudge. You can't analyze the working of that thing by linear mathematics because there's no cause and effect. You see, we're talking of circular logic now. All of scientific thinking during the last two centuries has tended to run on reductionism, which is Cartesian cause and effect, except that you can't analyze self-regulating things that way. They defeat the mind.

The same is true, of course, of quantum effects, like entanglement or light being both a particle and a wave at the same time. You just can't get your mind round that kind of thing. Sensible people like Richard Feynman, who was, I think, the greatest physicist in the last century, said, more or less, don't bother. Just accept that there's a lot of things you'll never, ever be able to understand. It doesn't really matter. You can use them in inventions quite easily, you can intuit them unconsciously, and you can write mathematical equations for them. That you can't understand them consciously is not really very important. That is true of Gaia and of a lot of things.

DC: You made a decision when you left the National Institute for Medical Research at Mill Hill to pursue an independent life. Are you happy that you did?

JL: Absolutely. I couldn't have made a wiser or a better choice, for me. I'm not recommending it for everyone, because not everyone has the same kind of outlook on life as I do. I'm very happy to work on my own. It suits me.

DC: How would you characterize the science that you've done?

JL: Well, it's quite easy. It took me a long time to realize it, but working in medicine for years helped. That is, I realize that I'm a GP in a world where there's nothing but specialists. That's what I am. As a scientist, I'm a generalist.

DC: You've operated on a small scale from your own modestly equipped laboratory, and yet you've made major discoveries.

JL: Ah, it's a great mistake to believe you've got to have billions in order to make discoveries in science. Somebody on welfare, some young person, would still have enough to buy a pencil and paper and to sit down and start thinking and come up with a really stunning theory. I'm sure it could be done. You don't need more than that. Einstein didn't have much more than that.

DC: What do you think has happened to science in your time — science as an institution?

JL: Too much money's been spent on it without enough regulation. You see, governments have felt — they're very tribal, are governments, and each nation wants to be leading. Our silly government felt, way back in the 1960s, if we pour money into science, we'll do wonders. We'll be producing geniuses by the hundreds. A moment's thought would have shown them that this is rubbish. If somebody else had said the same thing about art, that if we pour money into art schools we'll have Rembrandts by the dozen, they'd have said, what nonsense. They're born, not made. The same is true with science.

DC: When Jim Lovelock worked at the Institute for Medical Research in Mill Hill in the 1940s and 1950s, he published papers in *Nature*, the British scientific journal of record. The first paper he submitted to *Nature* after he went independent in 1961 was returned with a curt note reading, "We do not accept papers from private addresses." It's a telling instance of the institutionalization that he's just been talking about. But he has made the most of having had a "private address" within science. He has thought for himself, ignored disciplinary boundaries, and asked questions about things that everyone else seemed to take for granted.

In 2006 Lovelock published, I believe, his sixth book on Gaia, and it's by far the most alarming. It's called *The Revenge of Gaia*, and, in the first chapter, he likens the present generation of human beings to passengers on a small pleasure boat sailing quietly above Niagara Falls, not knowing that the engines are about to fail. Many current scientific forecasts try to blend caution with concern. They make conservative assumptions and are careful to hold out the hope that prompt political action can save the day. Lovelock's view is less optimistic. The reason he takes this more dire view, he says, is that he is one of the few who's looking at the whole picture.

JL: I've been privileged to look at the system through Gaian eyes, to see it as a whole system. Now, science, during the last two centuries before this one, has tended to be ever-dividing, getting into smaller and smaller fragments. I mean, on my last count there were about thirty different branches of biology and probably as many within physics and chemistry. Science is divided up and divided up, so it tends to see bits.

This came home to me during a visit my wife, Sandy, and I made to a famous climate centre not too far from where we are now, and it was this disturbing experience that led directly to the writing of that book. They were very good to us; they showed us all the work they were doing. Some were looking at the melting of the floating ice at the North Pole, showing us how quickly it's going and how within, what, twenty, thirty years, perhaps, you'll be able to take a sailboat to the North Pole instead of mushing there across the frozen ice. Others showed us the melting glaciers in Greenland and other places, and others showed us the way the tropical forests everywhere were beginning to go back to scrub and desert. We told them about our concern for the ocean and the way the top surface layer is warming so that the algae die. This is not because the algae are harmed directly by being warmed but because the top layer, when it warms, forms a stable layer into which water containing nutrients from below cannot reach. The algae just starve to death. That's why tropical oceans are so clear and blue. They're deserts, with very little life in them, whereas polar oceans look like soup because they're full of life.

Anyway, all of these different views of the earth were talked about separately. What alarmed us, by the time we got to the end of it, was that each of the groups we talked to looked at things in terms of their single problem. They knew about the others, because they all worked in the same institute, but they didn't somehow factor the work of the others into their own thinking. The glacier people talked about the tipping point that will be reached if temperatures rise 2.7 degrees. Then Greenland's glaciers will go irreversibly. There's no way of stopping them going. The forest people said, if the temperature rises by four degrees, there's no way the Amazon Forest can sustain itself, even if you do no forestry at all. And so on. They treated each bit separately. However, because of Gaia, I was forced to look at it from the whole-planet point of view. To me, as I said to Sandy as we drove back from the centre, this is the gloomiest story I've heard in years. It's terrible. That led me to write the book. I felt that when you looked at the whole picture, the earth was truly in — not so much in grave danger, because it's been through these things before, but civilization was in grave danger. The earth was now going to move from its rather comfortable cool state to a hot state that it's been in many times before.

The earth seems to have two stable states: cool, during glaciations and interglacials, or hot, as it was fifty-five million years ago during what's called the Eocene Thermal Maximum. That's its fevered state, if you like, but it's a stable state. That's where we're moving.

DC: A Gaian perspective is one reason why Lovelock thinks that the earth is inexorably moving to a hotter state. Another is the role he thinks positive feedbacks will play, and already are playing, in accelerating this change.

JL: The simplest example is the melting of the polar ice, the floating ice. Now, that ice, when it's there, is white, covered in snow, and reflects sunlight through the six months of polar summer straight back into space. The sun has very little heating effect on it at all. When the ice is all gone, there will just be dark black sea that will be absorbing almost all of the sunlight that comes in. As some of the ice melts, more black sea is exposed, which causes more heating, so more ice melts. That's positive feedback.

DC: In *The Revenge of Gaia*, Lovelock cites many other comparable forms of positive feedback that are likely to speed up change once it's begun. Putting the whole picture together, he concludes that the earth is about to turn inhospitable to the bulk of its human population.

JL: Good climatologists that I know, leading figures in the field — I won't mention names in order not to embarrass them — they tell me that by 2040 to 2060 — 2040, that's less than thirty-five years from now — every summer in Europe will be as hot as the summer of 2003, when twenty thousand people died of heatstroke and overheating. Of course, people might think, oh, well, by then they'll have air conditioning. It won't be so bad. But that's not the point. The point is, plants won't stand it. The agronomists tell me that if it gets that hot every summer, there'll be virtually zero agricultural production from Europe. So who feeds the 450 million?

When I say Europe, I'm excluding the British Isles and perhaps Denmark and Scandinavia. They'll be safe, although that just means, in effect, that all the other Europeans will have to move here or there. We're going to be pretty overcrowded, I reckon.

DC: Islands like the British Isles will remain habitable and arable because of the buffering effect of the surrounding ocean, but displaced populations will crowd in on them, Jim Lovelock predicts. And Gaia's capacity to self-regulate won't save us because it operates on an entirely different time scale than human civilization. Even so, the earth will recover.

JL: It's always recovered, and it's going to recover from this one, but it'll take two hundred thousand years. We won't have anywhere on the planet to live, except the polar basin, a few oases up on high land, and islands like the UK and Japan.

DC: Lovelock takes a long view of the earth's history. He knows the climate has changed before. One of the most extreme changes took place approximately fifty-five million years ago. Climate historians call it the Paleocene-Eocene Thermal Maximum, or PETM, and it occurred when atmospheric concentrations of carbon dioxide weren't all that much higher than the 380 parts per million that we have already reached.

JL: During the PETM, the CO_2 only rose to about 450 parts per million, according to Harry Elderfield, of Cambridge, who's the chief investigator of that period. The temperatures in the Arctic rose eight degrees Celsius. The Arctic Ocean was 23 degrees Celsius, and there were crocodiles living in it. I mean, there is good solid paleontological evidence of crocodiles up in the Arctic Ocean then. It was a warm place, but the rest of the earth was scrub and desert.

It's all very quick. It's coming on us so fast. That's why I'm very negative about sustainable development, renewable energy, all of these things. They're lovely ideas. If they'd been applied two hundred years ago, they might have done some good. Now they are hopelessly too late. Kyoto was very like Munich: a grand gesture, but not at the right moment.

DC: So what can be done?

JL: Oh, amelioration and adaptation are the two things that, really, we should be throwing our energies into.

DC: Lovelock's sense that we need to be thinking about managing the transition to a hotter earth, rather than building windmills, has made him a bit of a heretic amongst environmentalists. The worst of his heresy in their eyes has been his consistent support for nuclear power.

JL: Its use is to give a secure and, from a planetary point of view, non-polluting source of energy, and not even energy so much as electricity. Modern civilization won't work without electricity. Any major city would die in a week if you cut off the electricity and be reduced to a state like one of those camps in Darfur. It would be terrible. Work out for yourself what would happen if everything stopped — the lifts, the gas pumps, the water supply, the sewage system, the hospitals. Everything would stop. That's what would happen with no electricity, and that's why nuclear power is important. It gives you a secure, steady, reliable source of electricity that doesn't require any imports from a troubled world and doesn't pollute the atmosphere, as far as Gaia goes. I think people worry unnecessarily about nuclear waste, and I've offered to take the full output of a big power station here on this site. Personally, I'd use it for free home heating. It would seem a waste not to. Still, I understand that people have got an unreasoning fear of it, as they used to have of hellfire and various things like that.

DC: But you think that we have a lot of unreasoning fears, that we're so aware of small dangers . . .

JL: . . . and so neglectful of big ones. That's why I began my book, "Oh ye blind guides, you strain at gnats and swallow camels."

DC: It seemed natural to conclude my conversation with Jim Lovelock with a discussion of his latest book, but I was aware, in doing so, that I risked ending on a rather cheerless note. I asked him finally whether he had any words of comfort.

JL: Three upbeat things. Gaia will survive anything we do. Secondly, humans will survive. We're one of the toughest animals on earth, but we will be culled and, I hope, refined, so that among the survivors will be those who will make a better start for the next civilization. It's not that we have no chance. It's just that we're going through a very severe time of trial.

The other upbeat thing is that I'm quite old. I was a student at the beginning of World War II, and I remember the feelings before the war started. They were so similar to those now. That's why I compare Kyoto with Munich. There were liberal, good intentions everywhere about what could be done. The left said, let's disarm, and then Hitler will see us as no threat, and the right said, no, give him all the arms he needs, and then he will destroy the Communists, or the Communists and the Nazis will destroy each other in a big battle. It was negative, useless talk, but the moment the war actually started, everybody pulled together and made all the sacrifices necessary. They'd even sacrifice their lives. My goodness. The tribal forces pulled them together, and I do hope that the same thing happens again. Remembering my experience as a young man, I realize that it wasn't bad. It was amazing how cheerful people were in spite of their deprivation and how they found it a rather wonderful time to have survived. I think the same will happen again. We have already been through no less than seven events of this kind as humans. Humans have been on earth for one million years, roughly, as a species, and, during that time, there have been seven changes, from glaciations to interglacials, which are comparable in extent to the change in the opposite direction, upwards instead of downwards, which is occurring now. We've survived those changes, which must have been devastating. Just imagine that, fourteen thousand years ago, you lived in a small city civilization somewhere in Southeast Asia. What would you have thought of somebody like me who said, do you realize that in a relatively short time the sea level will be 120 metres higher?

Every Object,
Well Contemplated,
Changes Who You Are

ARTHUR ZAJONC

Goethe says, "Every object well-contemplated opens a new organ [of perception] in us." You have to live in that world of phenomena. You have to attend carefully. "Every object well-contemplated" — not just casually contemplated, but well contemplated, attended to over time, repeatedly — changes who you are, changes who you are to the point where you begin to see things that you didn't see originally, and perhaps which no one before you has seen.

— Arthur Zajonc

Johann Wolfgang von Goethe died nearly two centuries ago. Arthur Zajonc works at the cutting edge of contemporary quantum physics. But it is great German poet who Zajonc thinks can best show us how we ought to contemplate the puzzling discoveries of modern physics. Many of Goethe's literary contemporaries denounced the science of their time. The English poet John Keats called Isaac Newton's science "a cold philosophy" from whose "mere touch . . . all charms fly." It would, Keats said, "unweave a rainbow" or "clip an angel's wings." Goethe agreed with Keats, but he didn't stop at just criticizing Newton's philosophy. He wanted to show a different way of doing science, an alternative to the mechanical philosophy that would not unweave a rainbow. "The highest," Goethe wrote, "is to understand that all fact is really theory. The blue of the sky reveals to us the basic law of colour. Search nothing beyond the phenomena, they themselves are the theory."

Arthur Zajonc believes that Goethe's way of knowing points towards what he has hopefully named "the science of the future." We talked about Goethe when I interviewed him at his home near Amherst College in western Massachusetts, where he teaches physics. He told me first that he had been drawn to science as a boy and had believed it would provide an access to the truth of things, but then he was disenchanted.

ARTHUR ZAJONC

This longed-for insight which I thought science would provide for me — and, in particular, physics would provide, through the language of mathematics and experimental science — began to become, you might say, paler. I wasn't getting the full dimensionality, the full picture of what science had seemed to promise. You end up with a more and more abstract and a more and more remote understanding of the natural world around you. The equations begin to feel like they're interposing themselves between you and the natural world, as opposed to only elucidating it and getting you deeper.

While I was studying science and physics, I felt a great longing and, at the same time, a certain disappointment. That was a critical period in my maturation as a scientist, trying to sort out how to understand this science, which I felt had such promise, but which, in the end, was not delivering on that promise.

That launched me, you might say, into a study of the philosophy and the history of science, of the traditions out of which what I was studying in classroom settings actually emerged. When you're studying physics, you just get the latest stuff. You don't actually understand what the context was that produced this. I began reading quite widely in the history and philosophy of science, talking to professors outside of the narrow mainstream of physics instruction, and began to broaden my perspectives. That was an important watershed that took place when I was probably twenty-one or something of that sort.

DAVID CAYLEY

Arthur Zajonc's studies in the history of science put him in touch with inspiring figures such as Michael Faraday, the blacksmith's son who discovered electromagnetism and became, in Zajonc's words, the "greatest experimental scientist of all time." One of Faraday's colleagues, John Tyndall, described him as a man in whom the contemplation of nature produced a spiritual exaltation. "His religious feeling and his science could not be kept apart," Tyndall wrote. "There was an habitual overflow of the one into the other." The encounter with Faraday and others of his kind convinced Zajonc that there was more than one way of doing science.

AZ: You have two strands of scientific inquiry. One, you might say, is the Enlightenment strand, which is a kind of mechanical philosophy and an articulation of science in almost a redeemer role of providing insight and clarity. But there's a second strand which is, you might say, much more in contact with the material experience that Faraday was so excited by, the actual world of phenomena, the world of effects, the lived experience of doing science and discovering scientific insights. Ironically, I think, very often in science instruction, that second strand, the experiential strand, is de-emphasized, and the formal, analytical presentation is emphasized. As a consequence, one receives what I call the "pale version" — this somewhat denatured version of science without the experiential dimension. Every once in a while, there's a lab that you do, but it's pretty cookbook. There's not much in the way of actual originality. Even the phenomena themselves are diminutive. They're not very impressive. Whereas when you're actually looking back into the history of science, the moments of discovery are experiences: Isaac Newton's falling apple. He's looking at the moon. He sees the apple fall. He sees those two disparate phenomena as one effect, as one and the same. He stands before a set of phenomena, and he does the theory, but he does the theory in the original sense of the word "theory": to see or to behold. The formal, analytical treatment which we are taught, or which I teach to my students, comes much after the fact.

When people say science is often off-putting, it's too abstract, it's too hard, it's alienating, it's distancing, I can sympathize. That's something I actually lived through, and it took a lot of work for me to recontextualize, to bring the philosophical, the actual experiential dimensions back in. Often I had to go outside of the narrow mainstream of science. My studies of Goethe's science, for example, were very much part of that exploration. Here's a person who is known to the West as a poet, the author of *Faust* and other works, not as a scientist, but who spent much of his time, especially in his mature years, working with colour, biology, botany, and the like.

DC: Goethe's studies were carried on outside "the narrow mainstream of science." Hermann von Helmholtz, Goethe's countryman, was typical in his dismissal: Goethe, Helmholtz said, was a poet who focused only on nature's "beautiful show" and ignored the less glamorous backstage machinery that actually produces the show. Even so, Goethe himself remarked late in his life that he thought his scientific work was a greater legacy than his literary efforts. This work began with botanical research that he undertook when he was a minister in the government of the Duke of Weimar, but his scientific thinking really flowered during a trip to Italy in 1786, a trip undertaken to escape the cares of the court at Weimar.

AZ: Basically, Goethe runs away. He just leaves a letter for the duke. He departs under an assumed name at night, and he starts his journey over the Alps. When he comes to Italy, he's stunned by the landscape, and he's stunned especially by the differences in flora and fauna, the plants and animals, especially the plants that he knows from eastern Germany. The plant forms, which he's very familiar with from his own gardens, take on entirely new dimensions, shapes, sizes, textures, scents. It seems as if the context, which he has never really fully appreciated, provides an entirely new possibility for the development and metamorphosis of these plants. He begins to work most energetically at his "theory of metamorphosis" for plants, plant morphology, in this very different domain of experiences, this very different environment: more sun, more water, and so on. These plants flourish in ways he hasn't experienced before. He begins to write letters back to Germany concerning his understandings and his theory of plant metamorphosis.

He comes to a point — I think it's in the gardens of Palermo — where he experiences what he called the *Urpflanze*, the archetypal plant. In it, through all of these different metamorphoses and transformations, he begins to see a principle, as it were, which is eternal or somehow accessible as a defining nature of the plant world. From this experience, he feels he can predict all kinds of new species that are completely plausible, that is to say, which could exist in the world. He actually feels he can, in some ways, internally apprehend or see this, let's call it, "living principle" that stands behind the plant world.

Through a very meticulous study, first in Weimar, then in another whole environment in Italy, he develops a set of capacities in himself to apprehend — in the root sense of the word "theory" again, to see, to behold internally — a kind of living principle. It's not an abstract equation. It's not a genetic code that he's looking at, although some people have suggested he was intuiting something along those lines. No, no, I think it was much more a kind of internal apprehension of a principle which is alive in all the plant world.

That was the kind of work that Goethe did in the sciences. It wasn't work which was formal in the mathematical sense. He wasn't doing theory in the sense in which we use the word today. It was a working through of experience, often systematized through careful studies and sequences of what we would now call experiments, in both the physical sciences and the life sciences, that led him to an apprehension, what he once called an aperçu, by which he grasped that which is implicit as a living principle — in this case, in the plant world.

Then, at the same time he's there [in Italy], he's painting with a group of expatriate German painters, and, as he's painting, he has questions for them, quite interesting ones. He says, in a little essay called "Confessions of the Author," that he wanted to know why they used a particular colour to render the sky or the landscape around them. What determined their colour palette? They indicated that it was the *cognoscenti*, the art critics, that they were trying to please. They followed the style which was in vogue. This really annoyed him. Art should be done according to eternal principles, not according to what happens to be popular in the galleries down in Milan or some-place. This experience sent him on a chase, you could say, on a kind of inquiry, when he came back to Weimar. He came back with this burning question: what is the truth that's at the heart of the aesthetic use of colour? How is it that, if one were a true artist, one would use colour? He wasn't looking for a pedantic answer, but, as he wandered the ruins of Rome, he said, he could see there a principle which was as eternal and as true as any natural scientific principle. The arts were that objective, in a certain sense. This was an ancient culture and a very different period from his own — the art was very different from the art of his time — but he could see nonetheless that the principles that underlie all art were as true and eternal as the laws of nature. It was disconcerting to him, to say the least, that taste should be capricious, that it should just be what the critics like.

So he goes home, and he tells us that he pulls out an encyclopedia, opens it up to "colour," and finds the wave theory of light and the particle theory of light, and he realizes there's absolutely nothing here for an artist, nothing that's of any use. After some further explorations in the books that he has about him, he says, he was about to give up and just think it's hopeless, there's nothing in science that could benefit him or answer his questions. And then he realized that he had experience. He could refer to the world of colour himself. He could undertake an investigation himself simply by starting to look into the world of colour in a systematic and thorough enough way. Then he would come to insights, to the kind of living principles that are at the heart of that universe of colour, no longer the universe of plants and plant life, metamorphosis, but now whatever that world is, that universe of colour.

DC: Goethe carried on this investigation for more than twenty years, finally publishing his *Theory of Colours* in 1810. The book disputed many of the ideas of Isaac Newton, whose *Opticks* then completely dominated all discussion of the phenomena of light and colour. Goethe differed with Newton on many points. Newton held that light was corpuscular, composed of invisible particles. Goethe stuck to what he could see. Newton believed darkness was merely the absence of light. Goethe treated darkness as an active agent in the production of colour effects. But these factual differences are not what matter most to Arthur

Zajonc. What interests him above all is the two men's different stances. Newton looked behind the appearances while Goethe's approach was completely phenomenological. He described colours as "the deeds and sufferings of light."

AZ: They had very different ways of doing science. Newton was looking for a hypothesis concerning, you could say, the ultimate nature of light and the ultimate nature of colour. Goethe says, right at the beginning of his colour theory, that to expect that one can come to the true nature of anything abstractly — by that, he meant the way Newton was going about things, through theoretical discussions, selective experiments, and similar activities — is hopeless. Rather, what one should do is proceed by way of the history of the effects of light and colour. He gives an analogy. He says, to tell me about the nature of a person abstractly — and by that, I would say, something like what his IQ is, his height and weight, his personality . . .

DC: He's an introvert, an extrovert — that kind of description?

AZ: That kind of description . . . is a very pale and diminished description of that person. Tell me how he walks, his manner of speaking, the way he dresses, his interactions with others, and I immediately form a picture of his character. You can just feel the novelist or the playwright in him in that moment. A playwright can't, prior to the play or in the cast list, give the character types for all his characters. They simply appear on stage, and within the first five or ten minutes, if he's a competent playwright, each one of the characters is fully before you. You know who they are. You see into their souls, but you see into their souls through what they do, how they speak, what their mannerisms are, the way they dress, and all the rest. Goethe felt that this is the way one discovered the truth about the nature of reality, about the nature of the material and living world around us. It was not through an abstract description of personality types or something equivalent but through actually studying the effects, the performance of nature, and one encounters that through experiment, through observation. So one should do the same thing with this world of colour, and then you will infer, you will intuit, again, the kind of living principle that's at the heart of this domain of inquiry.

Now, Newton, by contrast, modelled his way of investigation on geometry and the proofs of geometry. His *Opticks* is set up in the format of axioms, postulates, theorems, and the like, and then he introduces certain kinds of experiments to support particulars of his theory. Goethe felt this was contrived. In other words, they weren't insights borne out of experience; they were hypotheses that were proposed prior to experience or on the basis of very limited experience, and then selected experiments were brought forward in support. One ended up with not so much an intuited core, a living principle, to which one was working through the phenomena themselves. Instead, one replaced that living principle by a model or what was called in those days a hypothesis. Then that model or hypothesis began to dominate. It took over your thinking. No longer did you experience a person as a person or nature as nature, you experienced them in terms of the models. The genetic code would be a contemporary example. All of the mind is to be encountered in the neuroscience of the brain. We have biomedical models and we have physical models and so forth today in abundance, and they become instantiated into the world. They become reified or made concrete. "Misplaced concreteness" is one of the great problems, Whitehead says concerning these models. We concretize them in ways which are problematic. The models themselves are innocent. As long as we have multiple models — and often contradictory models, as we discover we often need to have in things like quantum mechanics — we can see that the model is, in large part, an indication of our own mentality as much as it is a statement about the state of the world around us.

We fall in love with our models, and we practise then a kind of idolatry; the model becomes an idol. The idol, we forget, is just a pointer to something beyond, which is the living principle. We fall in love with the idol, and science becomes the practice of idolatry. Now, real scientists break through these idols again and again and again historically, so that's what you find in the history and philosophy of science: there's a realization, oh, my gosh, this is as much a picture of me as it is a picture of the world. Let me look at it differently. Let me get a different insight. Then a new model emerges, one which gives complementary insights into that same domain. And so the multiplication of models, even conflicting models, I think, is a great boon to science. The idea that you're going to find a single model which will somehow give an account of everything is hubris and, I think, a deception.

Goethe's style, you could say, the way his science differed from Newton's science, was that he rejected that kind of enterprise. What he does doesn't look like science in many ways, but if you look at the actual practice that he undertakes, it is faithful to the core principles of science, namely, it's empirically grounded, it proceeds from one methodical experience to the next, and it comes to a kind of insight, a moment of aperçu, of discovery. Goethe will not translate this into a mathematical form but will allow it to live as fresh experience. Then he will seek to apply it in various domains, in his own case very often in an artistic way.

Now, one last comment, because for me it's important: I think all good science — that is, original science — actually proceeds in the way Goethe describes . . .

DC: Including Newton?

AZ: . . . including Newton, in that, when Newton sees the apple's fall as the same as the moon's going overhead, he is seeing something. He's not writing down any equations. That comes later. He is living into the phenomena the way Goethe was living into the phenomena, and he's driven to it, interestingly enough, because there's a plague going on in Cambridge. He has to be back home with his mother in Lincolnshire, and while he's there, he has his two years of miraculous discovery — the so-called *anni mirabiles* — the two miraculous years in which everything happens. He is, in some sense, thrown back on his own resources, and he's thinking and observing, and he's pondering the questions of celestial motion, terrestrial motion, and their relationships. Thus he sees the celestial motion of the moon going overhead as the apple falling. To see the union of those two is an original insight. Then he develops the calculus in order to prove to himself that, yes, this is mathematically supportable. He creates all kinds of other methods of scientific investigation to support that, but the original insight takes place in a way that Goethe describes, I think, quite beautifully, the way that was, for Goethe, the heart of everything. That's what he was interested in. It was the artistic act, the creative act. He's not particularly interested in the explication, the theorization and so forth, of that insight.

DC: Mathematics is a way of recreating or remembering an experience. It's an important difference between Newton and Goethe that Newton had this ability; he invented a whole branch of mathematics, while Goethe did not. Zajonc knows mathematics, but he warns that mathematics can have its dangers.

AZ: There are two sides to the work that mathematics brings into science. Let's speak first about what you might call the positive and, perhaps, the beautiful side of what it brings. It brings on the one hand the kind of pristine clarity and lucidity which we all, all physicists, just delight in. It's just part of the pleasure of the discipline.

It also brings, surprisingly, a kind of unearned or undeserved power. By that I mean that sometimes you find yourself being led by the hand of mathematics further than you've gone yourself. I've been speaking about these insights, these aperçus, these moments of perception, and then about how you can mathematize those in part. However, sometimes, after having mathematized that insight, you find implications in the mathematics — something which you did not notice originally. In other words, the mathematics becomes generative. You begin to explore the mathematics, and you realize, oh, but there's another layer to this, one which I didn't notice phenomenologically. The power of the mathematics itself allows me to develop my insight further than I would have otherwise. This is probably nowhere more explicit than in quantum mechanics. Most eminent physicists agree that quantum mechanical systems defy understanding, in the conventional sense of understanding, in the way we normally understand physical systems. The eminent physicist Richard Feynman once said that people who think they understand quantum mechanics have rocks in their heads. He could do the mathematics — dead easy — but understanding quantum mechanics defied even his brilliance. In other words, mathematics provides for you a set of tools and methods which allows you to be, in some sense, more powerful, more insightful in the world than you actually are.

These are the two aspects of mathematics. On the one hand, mathematics can clarify and codify our insights and even allow us to explore and extend our understandings beyond our original insights. It leads us further. But there's another aspect which is also important to hold up: through the fact that mathematics is so centrally important in physics, it becomes the dominant form of modelling. You take a world which is complex, rich, textured, nuanced, infinitely contradictory, and you simplify and idealize and abstract that world into a form which is clear and lucid and unambiguous. In a certain sense, in that moment, you've also denatured it. You've taken away the multi-dimensionality of that world for a single dimension or for two dimensions. You can use those insights gained through clarification and simplification to great effect, but the danger, as with all models, is that they become idols. They become everything, and then everything is seen from that single viewpoint. It's that one monochromatic, blinkered eye that you see through. You see well, but you see only in one direction, one dimension, whereas nature provides itself with an infinite variety, infinite dimensionality. We have to be careful not to fall so in love with our own creation that we blind ourselves to all those other dimensions or think of them only as mere derivatives of that fundamental equation.

I think if one is self-conscious — and this is where the philosophy and history of science help us — if we're self-conscious about what it is we're doing, the tools we're using, the limitations of those tools, the fact that they don't merit being universalized and totalized, then the model is an aid to us. We should multiply the models again, find the ones that are contradictory, delight in the contradictions, and realize that the world is infinitely complex.

DC: Through his explorations in the history and philosophy of science, and in Goethe's work in particular, Arthur Zajonc restored his faith in science as a vivid experience of an inexhaustible reality. Models and formulas, he realized, stand between us and the world only when they become idols, only when we mistake the map for the territory. The contrast between Newton and Goethe was not the only thing drawing him to this conclusion. There was also the more recent history of his own field, physics. Beginning in the early twentieth century, physics had begun to reach into domains where no single model seemed to apply and philosophical puzzles proliferated. There was Einstein's theory of relativity, with its revelation that space and time are relationships and not independent realities. And then there was quantum mechanics, his specialty, in which matter itself seemed to decompose.

AZ: You could say that relativity theory has to do with undermining our conventional understandings of the space and time in which events and processes unfold, in which those objects have a life. Now, you say, let's look at the objects themselves, not just the space in which they happen to be moving around, which has already now gotten very interesting because of Einstein. Let's look at the objects themselves. What are they? What is an object? How does it come into existence? What is it made of? Well, what we know is that it's made of molecules, and the molecules are made of atoms, and the atoms are made of electrons, protons, and neutrons, and the protons and neutrons are made of quarks that are bound together by gluons, and so on, right? We tell a little story that goes all the way down. But then you ask, what are these fundamental constituents that we now have? Take the simplest example, the electron. It's a fundamental particle. We think it's not made of anything else. In other words, there's nothing like a sub-electron. So you ask the simplest of all questions: how big is it? After all, if the universe is made of spatially extended objects, things that have size, then the fundamental particles must be like bricks. You're going to stack them up, one on top of the other. An extended universe must be made of extended objects. So how big are the fundamental particles? Well, the answer is zero. They have no size. They have a location. They have a mass. They carry charge. Yet they have no size. They are point particles, as far as we can tell. Wrap your head around that. This great world of extended objects, you and me included, boils down to a set of things — for lack of a better word at this moment . . .

DC: . . . which aren't things at all . . .

AZ: . . . which aren't things at all, which aren't things at all. These point particles, though, have relationships to one another, that is, force relationships of attraction and repulsion, so they can configure themselves. Thus you have, as it were, nonentities which actually have attributes, attributes like mass and charge. They have no size, but they have a location. They are in relationships, but those attributes and relationships are also not simple. I'm still talking at this level about a kind of building-block universe, where the building blocks have gotten infinitely small, but now I have to explain that those properties which I said they have are no longer simple. They are quantum attributes, which means that that they don't have definite values. Think of an attribute. You have a certain height, you have a certain eye colour, you have a certain set of biometrics by which the immigration and naturalization people will be able to identify you when you come into the US. All those things are definites, and it's by those definites, that sequence of attributes, that we know each other.

But what if those attributes were ambiguous, not just ambiguous because I didn't know them, but fundamentally ambiguous? What if you had two heights, if you had two eye colours, if everything was in what we call in quantum physics "the coherent superposition state"? That is to say, it is both this and that. It's some kind of new relationship which is non-classical, which can't be thought of in a conventional way — it's not that you have one eye which is blue and one eye which is brown — but there's a kind of ambiguity concerning your eye colour such that, if I measure your eye colour, it comes up blue on one occasion and it comes up brown on another, but if I don't measure your eye colour, it has its own ambiguous — we say "superposed" — state, which I can make use of. It's not just ignorance concerning eye colour. It's actually a positive attribute. The ambiguity is a definite attribute or state of affairs which allows me to do certain experiments and which nowadays, with the advent of something called quantum computation, even allows me to build certain kinds of new machines which live off this ambiguity at the fundamental level, at the nature of substance. Not only have I disappeared the large-scale universe down to zero points, point-like particles, which are now in new kinds of relationships, but the very properties which we normally think of as inherent in these point-like particles are themselves in states which are quantum mechanical or ambiguous. Quantum mechanical and ambiguous, again, not in the sense that they're not known, but in the positive sense that our minds are not actually competent to understand them. Our mathematics are competent, our experiments are competent, we're driven to this conclusion, but to wrap our minds around this new state of affairs has proven essentially impossible. Niels Bohr felt it would always be impossible. When Feynman said that people who think they understand quantum mechanics have rocks in their heads, that's what he means. He says, we can do the math, we can do the experiments, we can build the quantum mechanical machines, we can sell them in the marketplace. Can we understand them in the way we understand the clockwork universe? No, because it is not clockwork. These new kinds of attributes, these quantum superposition states and so forth, require a new mentality.

DC: The fundamental ambiguity that Zajonc is talking about here is most easily illustrated by the nature of light. Light manifests as either a wave or a particle, depending on how the measurement is taken. Unmeasured light seems to have both characters at once — what he calls the coherent superposition state — and there simply is no unambiguous way to describe this state of affairs.

AZ: When we speak of light as having both a wave nature and a particle nature, these are two kinds of concepts. These are two concepts of the nature of reality — wave nature, particle nature — that are contradictory; you can't entertain them both at the same time. Something either has a particle character or it has a wave character. Each of them is well defined in its own right, but together they are contradictory. Niels Bohr called this feature of the new physics "complementarity." We are driven to have both concepts. We need both. These are the two models, if you will, in the language we were using before. These two models are both required, but they are contradictory, so we speak about them as complementary, and the principle of complementarity prevails.

DC: Quantum mechanics cannot circumscribe the reality it studies within a single description. It has to use multiple models, contradictory models. This inevitability of contradiction has led some physicists to say that we will never get to the bottom of things, but this is not Zajonc's view. He applies Goethe's maxim: "Every object well contemplated opens a new organ of perception in us." Yet isn't Goethe talking about the everyday world of visible, tangible things? Can his insight really be applied in the completely imperceptible realm of quantum effects? Zajonc's answer is a qualified yes.

AZ: Goethe is advocating for a phenomenology, and in quantum mechanics you're entering into a domain where essentially there are no phenomena, at least no phenomena that are visible to the eye or to the normal five senses, which was Goethe's whole locus. Goethe had particular reasons for being concerned with sensible things, aesthetic reasons. He was interested in the aesthetic use of colour, so he wanted to know what people's experience of colour was. It was no good for him to say, red corresponds to 600 nanometres. Can you experience 600 nanometres? What does it feel like to experience it? It doesn't feel like anything, so the measurement has no aesthetic value. He needed a form of exploration and a kind of science which was close to human experience. His questions were aesthetic and, in a certain sense, moral. He calls this combination *sittlich*, the "moral" use of colour, but it's both aesthetic and moral. By moral, he means to refer not so much to good and bad as to more emotive and affective responses — the psychological dimensions of colour. That was his orientation.

Now, you're moving, through technology, across a threshold. At some point, let's say in an experiment involving a single photon, you're beyond phenomena. You still have effects. These effects are registered by very specialized devices, gathered together over time in computers, and represented through mathematical methods and charts and graphs and the like; then we interpret them. But it's all by inference. These are two different worlds.

I kept thinking about this issue, and recently I wrote a book, together with a colleague, George Greenstein, called *The Quantum Challenge*. In that book, I basically give my response to the question, what would a Goethean quantum mechanical theory look like? How do you do quantum mechanics from a Goethean perspective? What I try to do in that book is to lay out a series of experiments. These are real experiments — they're not just thought experiments or abstract mathematics. Each one of them is designed to bring you as close as you can get to an experimental result which is, in Goethe's language, an archetypal phenomenon, except in this case there are no phenomena. You have archetypal *results*. You have to imagine your way across the threshold that divides the perceptible from the imperceptible. You no longer use direct perception, but the experimental apparatus is as simple as it can possibly be and still work. We've tried to clear away debris and minimize the inessentials so you just have the most clearly articulated, contemporary example of each of the primary concepts of quantum mechanics laid out. No one had ever done that. Previous presentations had all been more theoretical and cobbled together from thought experiments. The book has gotten quite a bit of appreciation because what teachers are able to do is lead students, step by step, experiment by experiment, through these archetypal moments. They're no longer archetypal phenomena in Goethe's classic sense, but they are archetypal experiments.

What does it mean? This is an interesting question. What is different for having sacrificed the direct experience of the phenomenon? When you look at a red colour or at a painting in a gallery, you have an inner response. There is a felt reaction, not only intellectual engagement but also a full, multi-sensory, internal response that is part of the phenomenal experience — very important. As you cross over into these more abstract realms of experience, that inner, lived response is diminished and replaced by a kind of pure intellectual response. I think that's an important threshold, a crossover.

You could say that many of our modern technologies do that for us. I grew up with my head under the hood of a car, fixing the engine. It's a lived experience. You actually get your hands dirty. You get grease on your fingertips, and the smells are there, and there's the joy of getting it to work and getting it to work better and faster and all that stuff. It's actually already different with electronics, especially modern electronics. It doesn't smell like much anymore. It doesn't feel like much anymore. There's no kind of visceral response. Increasingly, our technologies have that character. They provide us with services, but getting inside your cellphone is a hard thing to do. Whereas when I was twelve, my dad brought home a set of phones, and I took them apart and put them together, and I understood how they worked, basically. Now? Forget it.

We're in a space and time where we're surrounded increasingly by domains which are obscure to us, which are hidden from us, hidden not just because we're stupid but hidden from us almost absolutely. Perhaps what's lost there, and what's maybe of importance, is, again, the aesthetic and moral dimension. As long as you're coupled into the sensual domain, the aesthetic dimension, the moral dimension is also present with you. When you move across into the world of equations, abstract vector spaces and the like, which we inhabit when we're talking about quantum mechanics, the only aesthetics are the abstract aesthetics that apply to high mathematics. There's none of the sensuality. You're missing the aesthetic dimension that you have in the lived experience of colour, scent, sound, and so forth.

You can ask yourself, is anything endangered in that transition? If we inhabit a world which is that abstract, which is that disconnected from body, how do we make moral judgments? How do we make aesthetic judgments unless we import them from another domain of life, unless we bring them along with us? I think some of that comes up with the Manhattan Project and with the genetic technologies that we currently have. They're disconnected from lived experience. How do you live your way into the calculations that stand behind the atomic bomb until the thing goes off? Then, all of a sudden, you realize what you did. In my view, we confront that increasingly: our technologies, our insights outstrip our moral development in part because we've disconnected our technologies and insights from the body, from sensual experience. It's a characteristic of the new science. It started in physics, but it's making its way into molecular biology and now into neuroscience. As we increasingly rarefy and make more abstract these domains of discourse, these domains of exploration, they become disjunct, disconnected from the normal grounds by which we judge the aesthetic and moral dimensions of life. Consequently, we need to re-embed those discoveries somehow back into our lives. But we need to do so consciously. It's not just part of our nature. It doesn't just come naturally anymore because we're beyond nature in a certain sense. We've extended nature into a domain where our normal equipment isn't sufficient.

I was asking before, do we have the concepts to understand relativity and quantum mechanics? One can ask as well, do we have the moral competencies to handle them? It's not just our intellectual concepts that are at a loss, that we're having to retool. We're also trying to develop moral capacities and aesthetic capacities that can, in some sense, re-embed these new dimensions of science that emerged first in the discoveries of quantum mechanics, relativity, atomic weaponry, and so forth and that now are equally advanced in molecular biology, genetics, and neuroscience.

DC: New technologies and new realms of scientific exploration take us beyond the body and into domains where we lack any moral or aesthetic grounding. This is why Zajonc keeps returning to Goethe. Quantum mechanics can make us wonderfully aware of certain deficiencies in our understanding, but only contemplation can make us whole.

AZ: I come back to Goethe. To me, these two were knit together in my biography. Quantum mechanics does one whole piece of the work — it raises wonderful puzzles, it points out many of the shortcomings in our classical understanding — but it doesn't do all the work. We can't rely on it to do everything. The other piece of the work, which Goethe brings, is a return, again and again, to human experience and the possibility of developing that experience beyond our current horizon. Our current horizon, that is to say, is nothing more than the limitations of our own capacities, and, he says, "Contemplate well and new capacities will open." If we are in a new terrain, a terrain which is posing these deep mysteries concerning the ultimate nature of reality, pay attention, contemplate well.

Another part of my life has been that world of contemplating well — the contemplative life. Parallel with the time that I'm talking about when I was studying Goethe, I was also reading in the mystical and spiritual traditions. Goethe's language, his language of contemplating well, resonates, to me, with the best of those traditions, which are not really traditions about metaphysics so much as about experience. I think we can leave aside our pre-commitments due to religious traditions. As valuable and as important as they may be individually, they tend to separate us.

Our traditions have their place, and their honourable place, in world history, whether we're Christian or Jewish or Muslim or Buddhist or whatever. But we can also proceed scientifically, and science, at root, is grounded completely in experience. It's an empirical endeavour. Is it possible to broaden the range of that experience, and, if so, how? The methods I have found most valuable in that regard are the contemplative methods. Increasingly in my teaching, and in my writing and so forth, I weave those dimensions into my work. My work with the Dalai Lama in the *Mind and Life Dialogues* joins both of those strands. We're really working hard to look at the nature of reality from both a scientific standpoint and a Buddhist, philosophical standpoint. As the Dalai Lama says — and I agree with this — the more understanding we have concerning our own nature and the nature of the world in which we live, the better is our chance of mitigating suffering, because often our actions are based on delusions, are based on reifications that arise from our own thinking, our own very narrow-minded and parochial thinking. If we can get outside the box, get outside of that thinking and begin to appreciate the context in which we have been living up until now and change that context, change the way we think, then we can reduce the attachments, we can reduce the suffering that we endure ourselves and that we inflict on others. This is not only an intellectual enterprise. It's actually an act of loving compassion to become clear, to become honest about the world in which we are.

It's not good enough to just work your way through these problems intellectually. You have to change who you are, and the way you change who you are is through contemplation, through contemplative practice. I've been working for ten years or so with several hundred or a thousand other faculty in universities around the United States and Canada to bring that kind of message to students and to say, listen. On the one side, there is a pedagogy of information, where you need to know certain things. On the other side there's a pedagogy of transformation, and that transformation takes place at the hand of reflection, takes place at the hand of contemplating well. What are the methods? Some of the methods are drawn out of Goethe and out of William James and his philosophy of experience and radical empiricism. Some of them are drawn out of the contemplative traditions of Asia and the West. Buddhism and [Rudolf] Steiner are two main sources for me. These are traditions which I think can be enfolded into a sensible pedagogy for young adults and beyond, and which ultimately extend our horizon and do so in a way which is honouring the scientific tradition, which has rooted itself throughout in experience and reason. We want the same in this more expansive understanding of our own capacities. We can expand those capacities through reflection and contemplation so that experience is broadened, and we can bring reason, not only reason which is familiar to us, but maybe even a new kind of reason which can take up these quantum challenges, take up the new experiences that we gain at the hand of introspection and reflection. It's a new kind of thinking joined also to an enlarged domain of experience.

What Needs
to Be Subtracted

WENDELL BERRY

*The standard of science must be nature, insofar as nature contains us,
comprehends us, and ultimately judges our behaviour.*
— Wendell Berry

Wendell Berry is known to the reading public mainly for his poems, essays, and novels, not his commentaries on science. However, in the year 2000 he published a surprising book called *Life Is a Miracle: An Essay Against Modern Superstition*. The superstition the book denounces is the belief that science will one day give us a complete account of things. Science is admirable, he argues, but it can only be deployed wisely when we recognize the limits to our knowledge. Science must submit to the judgment of nature.

Berry's first encounter with the misuse of science was in agriculture. For more than forty years, along with his writing, he has worked a hill farm in the part of Kentucky where he was raised and where his family has farmed for generations. During that time, he has watched the end stages of what he has named "the unsettling of America" — larger, less diversified farms, fewer, more indebted farmers, degradation of land, rural communities broken and scattered. And he has seen the role agricultural science has played in fostering the mentality and the techniques that have produced these results.

Meanwhile, he has been moving in the opposite direction, towards the restoration of community and the conservation of land. Two of his great teachers in the proper use of land were Sir Albert Howard, a pioneer of soil conservation and organic agriculture in the first half of the twentieth century, and Berry's friend and contemporary Wes Jackson, whose Land Institute in Salina, Kansas, is dedicated to the development of a sustainable prairie agriculture. Both insist that nature is the standard against which agricultural science must be judged. Wendell Berry spoke to me from a radio studio in Louisville, Kentucky, and I asked him what it means to judge science against nature.

WENDELL BERRY

To accept nature as a standard is to accept the local ecosystem and its health as the chief indicator of the health of the human economy in that place. Sir Albert Howard said, for instance, that if you're farming in a country that was originally forested, then your exemplar, your pattern is given to you by the forest. Wes Jackson has said, because he comes from Kansas, a prairie state, that if you're farming in a country that was originally native, tall-grass prairie, the native prairie gives you your example and your standard. You must perform in your use of the land the same functions of land preservation and diversity that the natural ecosystem performed before you got there.

DAVID CAYLEY

How does that apply to your own farming? You farm on formerly forested land. How does the forest give you your standard?

WB: Well, I need to tell you that I farm on very marginal land that is mostly — nearly all — either in woods or in grass. And it's steep land. What level land there is is marginal because it floods periodically. So I have a farm from which you can learn a great deal in a hurry. The first lesson to anybody farming steep land is that you've got to keep the ground covered. Sir Albert Howard said nature always keeps the land covered, always farms with animals, maintains a great diversity of species, maintains large reserves of fertility, wastes nothing, conserves the water, and so on.

DC: Farming methods, in Berry's view, have to be tailored to the particular places where they're used. One of his main criticisms of conventional agricultural science is its indifference to local peculiarities, and one of the things he admires in the work of his friend Wes Jackson is his sensitivity to place.

WB: What keeps me an enthusiastic friend of Wes Jackson's Land Institute is its absolute acceptance of its local and ecological context. It knows that to find the right way to farm a prairie is not to find the right way to farm a valley in the mountains of Kentucky or a coastal plain in California. In other words, what is admirable about that kind of science is its courtesy, first toward the place where it studies and applies itself, and second, toward mystery, toward what it doesn't know — its acceptance of its own ignorance as a premise.

DC: Berry has been recommending ignorance as a premise ever since he published *The Unsettling of America* thirty ago. His latest collection of essays is called *The Way of Ignorance*, but he intends no depreciation of science. He is calling, rather, for a science aware of its limitations and its need for controls.

WB: Science for me is a way of knowing a kind of truth fairly certainly. In that way it's admirable, but by the time I had finished *The Unsettling of America*, I had seen the disruptiveness of industrial agriculture. I saw that it disrupted both the landscape and the human community. I didn't think agricultural science, as it was being practised, was comprehensive enough. There weren't the necessary control plots, for instance. What we were doing in agriculture was not being measured against natural ecosystems or against better kinds of agriculture. In other words, industrial agriculture was being carried on without experimental controls.

DC: Industrial agriculture developed without any standard of comparison. It didn't compare its fields to uncultivated land of the same type or to lands under other forms of cultivation. The problem, for Berry, wasn't that it was unscientific but that it wasn't scientific enough.

WB: Industrial agriculture followed a logic of application of industrial methodologies to the landscape. The science that served it was simply working out that logic — without controls and very largely without examination of results. The disruption of farming communities, for instance, or the damage to the economies of farm families simply didn't enter into the calculations, nor did the ecological effects of the use of toxic substances, or the ploughing out of fence rows, or the substitution of annuals for perennials, or the reduction of land in pasture. The results were not calculated or considered, as far as I could tell.

DC: Why do you think they were not considered?

WB: Because the logic of industrial land use was enormously profitable to the agribusiness corporations. They had the advantage of spectacular increases in production to show for what they were doing. But their standard was simply the standard of production and, of course, profit.

DC: What about the model of science that was being employed — the way of knowing?

WB: The way of knowing simply was to develop a remedy or a solution that was a techno-scientific industrial solution and then apply it. If it worked to increase production or control weeds or control insects, then it was applied — and applied universally — in agriculture very quickly. The overarching problem was in this universality of application. That is, it was assumed that what worked in one place would work in another. From the beginning of industrialization, what we had going on in agriculture was the opposite of local adaptation, with, I think, dire ecological and social results.

DC: Is this inability to take local circumstances into account inherent in the scientific project? I don't mean in its ideal form but in the way it has in fact been structured in the West.

WB: It's inherent in a professionalized science and a university system that does not identify with any place. The people involved don't register the effects of what they do by examining the results in any place that is dear to them. My perspective as a critic has been that of a person who identified absolutely with a local place and a local community. That's the difference between my work and theirs.

DC: During the last century, universities have been the main incubators for the techniques used in industrial agriculture, but they have been removed from the consequences of these techniques. This abstracted science, Berry writes in *The Unsettling of America*, "is saved from the necessity of killing the bearers of bad news because it lives at the centre of a maze in which the bearers of bad news are lost before they can arrive." No particular place ever has to be considered.

WB: Research in the modern university is understood as inherently good, and it's very hard to discover if any standard of judgment is being applied anywhere in the system as to its possible effects on the landscape and on the farming communities. The question ought to be, is this healthful? Is it healthful to the landscapes? Is it healthful to the farmers who are performing the application and their families and communities? And finally, is it healthful to the consumers who eat the food thus produced? I think health is the inescapable standard, and it's the standard by which this work ultimately will have to be judged. I don't think the question about healthfulness, which would have to be a question locally applied, is being applied.

DC: Health is an attribute of a person or a place as a whole. As Berry points out in *The Unsettling of America*, the very words "health" and "whole" come from the same Old English root, *hal*. Heal, hallow, and holy also share this descent. However, the whole is no one's responsibility. Modern sciences have generally been reductive, breaking things down into smaller, more easily managed parts. The result has been an ever-growing specialization. Not always a bad thing, Berry says, but deadly when carried too far.

WB: I don't think that specialization is inherently wrong or damaging. If you want the best work to be done, you have to have that work done by experienced people, and this means you have to have a measure of specialization. If you want a new roof on your house, you'd prefer to have it done by somebody who'd put on a roof before. Specialization is, in that sense, practical, but if the specialist is specialized to such a degree that he or she can't speak to colleagues in another field, or can't look as a citizen and a neighbour at his or her speciality and judge it according to what it is doing to people and to places, then that's too much specialization. It oughtn't to stop people from asking responsible questions that arise from citizenship and local loyalty.

DC: Local loyalties have been badly battered during an era when the conventional wisdom in agriculture has been, grow or die, get big or get out. The closely knit, largely self-sufficient farming community that Berry has chronicled in his novels and stories is mostly gone. Yet in the midst of this destruction there has also been a revival.

WB: I think there's a growing number of people all over the world, really, who are refusing to accept the necessity of professional indifference and overspecialization. They are making their refusal concrete by supporting local farmers' markets, community-supported agriculture, local food economies, and that sort of thing. I live in a county that still is predominantly rural, with a comparatively small population. It has its problems. The land is being priced out of the reach of farmers, for one thing. But there is a farmer's market in the county seat, a small town. The farmer's market takes place in the courthouse yard on Saturdays and on Wednesdays. It is well supplied with local food and well patronized by local people. My wife and I are friends also with a pair of young people who have a very successful community-supported agriculture farm. These things are going on, and they're going on nearly everywhere I go.

DC: Berry's first meditations on science were in connection with the agricultural themes that he has just been addressing, but in 2000 he widened his focus with a book called *Life Is a Miracle: An Essay Against*

Modern Superstition. Its immediate occasion was a book published two years earlier by Edward O. Wilson. E.O. Wilson is a biologist, an outspoken conservationist, and the inventor of sociobiology, the name he gave to an ambitious book, published in 1975, that attempts, in his own words, a "systematic study of the biological basis of all social behaviour." In 1998, Wilson published *Consilience: The Unity of Knowledge*, the work that attracted Wendell Berry's attention. It carried forward the grand synthesis he had begun with his *Sociobiology* by proposing that all knowledge should one day be unified under the aegis of science. Berry found the argument alarming.

WB: The book frightened me. I think that book is typical of a certain kind of scientific statement that is coming along pretty frequently these days in which a scientist of stature will depart from empirical science, that is, the admirable ability of science to establish certain kinds of truth, and venture into prophecy and into advocacy of a triumph of scientific methodologies that would *dominate* life. That, to me, is terrifying. When I read *Consilience*, I began to ask myself what I had in myself and in my mind and in my reading that could oppose what I considered to be the threatening impulse of Mr. Wilson's book.

DC: The gist of *Consilience*, which literally means jumping together, is that all the elements of human culture should eventually be brought under one unified description, and this description should be modelled on the kind of explanation that has already succeeded in the natural sciences. Wilson puts forward his ideas in a spare, elegant, and humane style, but, in the end, he makes no bones about his claim that religion and art and human psychology will all be shown to have a material and biological basis.

WB: The book seemed to me to be frighteningly reductive. I think that human dignity, for instance, depends on not knowing everything. You and I have a measure of dignity and are granted a measure of dignity because just anybody can't presume to understand us entirely. That dignity, that mystery, seems to me to be infinitely worth protecting.

DC: Do you feel it to be particularly threatened in our contemporary circumstances?

WB: Yes, I do.

DC: By the advances of science?

WB: By the advances of science. By the willingness, for example, of agricultural science to apply the products of genetic engineering on a huge scale without adequate debate or experimentation or foresight or respect for local conditions and the integrity of local ecosystems. Yes, I think that's threatening. And political power tends to operate in the same way. I am for a kind of scientific and political forbearance that allows a just and reasonable measure of individual and local self-determination.

DC: E.O. Wilson's *Consilience* presents a progressive image of science. Science is moving towards a possible but yet-to-be-achieved unity of knowledge. The best, in a sense, is yet to come. This is one of the elements of the book with which Berry engages most vigorously. In his view, the idea that revelation is just around the next corner or just over the next hill has two grave and pernicious consequences. The first is that it runs together what we don't know yet with what we'll never know. Second, it treats the past as no more than a sketch of a better future.

WB: One of the most interesting characteristics of Edward Wilson's book, to me, is the prevalence throughout of the frontier metaphor. This seems to me to be a kind of romanticism. I don't know whether science inherited it from the arts or exactly where it comes from, but we hear all the time about the science of the cutting edge. The emphasis is on novelty, and the apparent assumption is that the business of human beings is to be always going where they've never been before. So we hear about the frontier of science. We hear about the virgin land and the necessity to go there. Scientists of the wrong kind, it seems to me, are imagining themselves as the Columbuses and the Daniel Boones of our era. This seems extraordinarily dangerous because, for one thing, it removes value from all things past and places that uprooted value entirely upon the new. Too often, it's the untested new that receives this high evaluation. We tend to denigrate the achievements and the thoughts of our forebears, and this means that we lose our cultural inheritance as a guide.

DC: E.O. Wilson, as Berry reads him, sees scientific knowledge as a steady advance. Whether science really is progressive in this way is a question that continues to exercise philosophers of science, but it's somewhat beside Berry's point, I think. What he's trying to draw attention to is the damage done by what he calls the metaphor of the frontier, and especially to the shadow this metaphor throws on the past. This is one of a number of examples he develops in *Life Is a Miracle* of the harmful effects that occur when science exceeds its proper boundaries and, so to speak, overflows its banks. Another is the damage that is done to everyday talk when it is colonized by scientific abstractions.

WB: One of the effects of science is that it's given us an undue love for mouth-filling polysyllables, which means that we now very readily speak a kind of gobbledygook using a vocabulary that we only partially understand. Medical science has taught us to do this. Almost everybody now is talking about drugs and medical procedures and so on that they really don't have any useful understanding of at all.

DC: What happens to the capacity to speak to one another when our discourses are full of these half-digested migrants from science?

WB: I think it reduces our ability to speak particularly to each other of the things that particularly concern us, but it also involves a transfer of authority from personal experience, local experience, to general knowledge and general authority. For instance, if I learn to farm my place from official or commercial agriculture, I'm going to make serious mistakes. If, finally, I farm my place well and use it conservingly, I'm going to have to depend on my own experience there and my own knowledge of it. The same is true of my relationship to my wife and children. There's a limit to how successfully I can relate to them by means of general psychological or scientific advice. Finally, I'm going to have to know the person I'm speaking to, and, worst of all, I'm going to have to know myself and recognize my own weaknesses and my proclivity to always be right.

DC: How does that apply to language, then?

WB: It means that the language we use has to have the ability to particularize, to say precisely what we mean about the things we uniquely know. It's terribly insulting, for instance, to have somebody presume to understand you completely, especially if that person is, say, a doctor who doesn't know you very well at all. I think a lot of people who have been medical patients understand this: your encounter with your medical expert becomes a kind of contest to see if you can assert and defend your particular problem and your particular self against the medical generalization that's being applied to you.

DC: Science inevitably generalizes; experience is unique. This becomes a problem, Berry argues, only when the difference goes unrecognized, when, as in his last example, experience has to defend itself against scientific generalizations. Again, the problem is not science as such, but science unaware of its limits and possible perverse consequences. Knowledge can be a mixed blessing, intensifying the darkness even as it increases the light.

WB: As knowledge increases, problems increase with it. If your scientific knowledge grows without due attention to the proper use of that knowledge, according to adequate standards, then you can say that, as the radius of knowledge lengthens, the circumference of ignorance proportionately increases. One way to get at this is to say that our primary practical obligation as a species is local adaptation. Most modern scientists accept evolution and the accompanying truth that every species in its evolution has come under the necessity of local adaptation. If an animal or other organism fails to adapt to its place, it fails altogether. A continuing wonder to me is how these people have managed to exempt our species from this requirement. It is, I think, permissible to say that as our general knowledge and the power of that knowledge increases, the power to do damage increases, insofar as it fails to accept this obligation to adapt locally.

DC: Some years ago, a book series called American Authors devoted a volume to Wendell Berry. Among the many tributes and critical essays was a letter from a writer who had been Berry's teacher at Stanford in the 1950s, the late Wallace Stegner. "By every stereotypical rule of the twentieth century," Stegner wrote to Berry, "you should be dull." A few of the modern rules Stegner had in mind are that the city holds more interest than the country, the future than the past, evil than good, leisure than work. Wendell Berry is far from dull, as Wallace Stegner knew, but as an artist he has reversed all those rules. His subject in his stories and novels has been the goodness of a vanished way of living and working in a tiny region of rural Kentucky. One of the things he celebrates most vividly in these writings is the art of farming as it was still practised in his childhood. He writes of the balance that existed in the old mixed farms between men's and women's work, between cash crops and subsistence, between cultivated and wild land, and among the many elements of the farm's internal economy. He writes of the extraordinary range of competencies that this way of life fostered and of its arduous but unhurried pace, in which there was still time for local talk and space for the flowering of character. This style of farming, except for the occasional holdout like Berry himself, was largely destroyed by industrial agriculture. Farms became dependent on methods developed in distant universities and products purveyed by huge corporations. Farmers lost their time and their autonomy to a debt treadmill. They became consumers.

WB: Increasingly in industrial agriculture, the consumer of the methodology or the products has been a passive follower of highly generalized instructions, so that, as the uniformity, the abstracting power of this technology takes over, one of the critical losses is intelligence. In other words, the consumer or the user permits the intelligence concentrated in one corporation or one central place to replace the intelligence of many individuals. I think that if we take the possibility of good land use seriously, and if we take the necessity of local adaptation seriously, we have to see that we need the intelligence of every mind that's involved. We need that intelligence to be strenuously applied. We don't need the implied requirement that all people simply submit to the intelligence of a few selected and hired minds somewhere else.

DC: Industrial agriculture centralizes intelligence. The farmer, reliant on standard methods and standard inputs, loses his feel for his particular landscape. This might seem a romantic or nostalgic judgment, but I don't think that it is. Berry knows, and has written, that long before his time, American farming was already divided between what Wallace Stegner called the boomers and the stickers — the boomers interested in exploiting and moving on, the stickers committed to stewardship and sticking around. Berry recognizes that land was abused in his own part of Kentucky well before the farmers' peonage to agribusiness began. At the same time, he does hold that the independent farmer is more likely to use the land well. We take care of what is our own. Property and propriety are essentially the same word — doublets, my dictionary calls them — and appropriate belongs to the same family. Propriety is a big theme of *Life Is a Miracle*. Years ago, in a book called *That Hideous Strength*, C.S. Lewis suggested that in the course of scientific progress we learn "to stifle all deep-set repugnances." Berry, too, thinks that, in our rush to open all existence to the scientific gaze, we override certain decencies, a certain tactfulness, that belong to our feeling for what is ours.

WB: Decency and tact probably depend on our ancient sense of what is appropriate. We have the word propriety, which we've mostly discarded, because we associate it with a kind of stuffiness, a kind of meaningless etiquette about how to behave. But propriety really means the ability to do what is appropriate in given circumstances. By saying that, we re-energize the word and give it a practical force. We would say that decency in our social relations consists in our ability to do and say what is appropriate in certain circumstances. If we're in a house where there is a newly dead parent, there are certain things that it would be inappropriate to say, and decency requires that we don't say them. If we're in a house where there's a seriously ill or afflicted child, decency requires us to know what is appropriate to say in that circumstance. Well, decency in land use implies and requires the same sort of propriety. We need to know where we are, what the circumstances are, and then we can maybe proceed to a reliable idea of what is appropriate to do in that place. The question, the practical question, is, what can we do there without doing damage that might be so great we can't repair it?

DC: Do you think that a certain kind of science has injured those sensibilities?

WB: Well, there's no question in my mind about that. If we are scientists, and if we assume that we can simply pursue our speciality in a laboratory or a university and that our work will inevitably have good results, then we're involved, it seems to me, in great personal confusion and possibly in great danger. The modern history of science alerts us to that possibility. If you pursue nuclear physics, for instance, with the assumption that your only obligation is to your science, then you may have to deal ultimately with tragic and disturbing results. We know that that has happened. I think the same thing is true of other sciences that become interesting to wealth and power. We can't simply throw it out there with the assumption that some kind of benevolent whirlwind will direct it only to desirable ends. So yes, it's possible to pursue science in a destructive way, but then, of course, it's possible to pursue any discipline in a destructive way.

DC: One of the things that it was interesting to find in *Life Is a Miracle* is the little digression, or perhaps it's not a digression, on obscenity in literature, where, again, there's the sense of frontier, of the edge always having to be pushed. This seems to me to be part of a larger question about how science overflows into the arts and into the humanities.

WB: Yes. Wes Jackson calls this physics envy — the wish in the humanities for the stature of the frontiersman, the discoverer of virgin territory, the first to enter the unexplored place — and this sometimes has inspired people with the wish to speak what has heretofore been unspeakable. It seems to me that there we very quickly run into a limit. Obscenity has a certain power, especially to correct over-refinement. I understand that, and I'm somewhat drawn to the obscene and the vulgar myself, for that reason and also because it can sometimes be funny. However, there is a limit to that. In that book, I talked about the willingness on the part of certain writers to exploit their knowledge of their kin and friends for the sake of art. My own feeling is that art is not worth ruining a friendship or a family relationship for. There are certain things you shouldn't do to your friends and family. Or to anybody. You shouldn't embarrass somebody, for instance, for the sake of your art.

DC: You called *Life Is a Miracle* "an essay against modern superstition," which is a bold thing to say about science. What is the precise sense of superstition there?

WB: Essentially, what I mean by superstition, by modern superstition, is the assumption that our knowledge will ever be complete or even entirely adequate to our needs and our experience in this world. We are going to continue to be surprised. Nobody predicted the destruction of New Orleans, for instance. Nobody, in spite of our much appreciated and praised intelligence, anticipated the catastrophe of 9/11. One of the common human experiences is the experience of surprise, and even the smartest people are subject to that experience.

DC: Modern science was conceived as a project of prediction and control, but few things turn out as we intend, and so we are endlessly surprised. The less we reckon on surprise, the nastier the surprises turn out to be. Berry's proposal is that we begin to keep a more truthful balance sheet in which we show our losses as well as our gains. Modern medicine gives him his example.

WB: Well, of course, the history of modern medical science is a spectacular and dramatic history because it's the history of a series of cures, of immunizations and so on. I think we've extrapolated from this history of successes to what I would call the superstitious assumptions that we will eventually be completely successful and that death and suffering in this world can therefore be looked upon as abnormalities. Again, I think that we're going to be surprised. I think we're going to find that mortality and suffering are, beyond some point, not reducible, and that wisdom and the ability to live well are going to depend on our ability to accept and deal with suffering and death. These are going to be hard things to deal with. They always have been. Nobody deals with them except by having to. But we have the cultural resources for dealing well with these things, and it's a terrible mistake to depreciate those resources or disregard them.

For instance, I was at a meeting not very long ago at which I heard a young doctor say confidently that the efficacy of modern medicine was proved by the increase of longevity. Now, my contention is that our great cultural debility is our inability to subtract. In other words, we have this uncanny and not very admirable ability of carrying on as if every innovation was a one-hundred-per cent net improvement. The young doctor who was praising our ability to increase longevity was not deducting the misery of people who live too long. The statistics on longevity, for instance, include people who are living, in effect, without minds and without appreciable body comfort and certainly without much pleasure in nursing homes. Those statistics include people who have been seriously damaged by the medical industry itself. Before crediting or passively consuming such claims, we need to see what needs to be subtracted. The net good, I think, will be smaller than is claimed.

A Heretic in the
Church of Science

*Biology's attitude to animals and plants seemed to be: kill them first,
and then ask the questions. It didn't seem to be about life. It seemed
to be about death.*

— Rupert Sheldrake

Rupert Sheldrake recoiled from a scientific education in which the only
way to learn about things involved cutting them up into smaller and
smaller pieces. He wanted what he called *A New Science of Life*, the
name he gave his first book, published in 1981. It asked a question
which had long vexed modern biology: how do organisms develop into
the intricate and involved forms that we see around us if all they have to
go on is the sketchy information contained in their genomes? How does
an acorn actually become an oak? The standard answer to this question
was, and remains: it's all genetically programmed. But Sheldrake wasn't
convinced. He put forward an alternate theory: that there must be some
kind of form-giving field that holds the memory of each thing's proper
shape. He called it a "morphogenetic field." It's an intriguing idea, and it
was widely discussed for a few months after Sheldrake's book appeared.
Then Sir John Maddox, the editor of the prestigious scientific journal

Nature, wrote an editorial which violently denounced Sheldrake's work and called it the "best candidate for burning there has been for many years." Maddox, in effect, excommunicated Sheldrake, and, when Maddox was interviewed about this famous editorial for British television many years later, he made this quite explicit. "Sheldrake deserved to be condemned," he said, "for exactly the same reasons the Pope condemned Galileo: it's heresy."

This attack on the theory of morphogenetic fields gave Rupert Sheldrake a reputation for flakiness that still lingers. A few years ago, Nobel physicist Steven Weinberg was still referring to the theory as a "crackpot fantasy." For Sheldrake, though, this zealous policing of the boundaries of science only proved that scientific materialism had hardened into a rigid and inhibiting dogmatism, and he carried on with the research program he had put forward in *A New Science of Life*. I visited him at his home in London, and he began by telling me a little about his early life. Like many scientists, he was first a naturalist, fascinated by living things. In one of his books, he recalls his amazement as a boy at learning from an uncle that a row of willow trees near his home had spontaneously regenerated from an old willow fence. This keen interest in nature was fostered by his father.

RUPERT SHELDRAKE

My father was a herbalist and a pharmacist and a microscopist. The next room to my bedroom had microscopes in it, big, brass microscopes with glass domes over them, and boxes full of slides and devices for taking pictures of things down microscopes. He showed me a lot of things under the microscope: little creatures in pond water, bees' tongues, the scales on butterflies' wings, cross-sections of plants, and other things. This was a wonderful introduction to science for me. Really, he was a kind of naturalist. He knew about plants. He collected plants. I had a pressed plant collection. Biology was something that I knew I wanted to do from a really early age.

DAVID CAYLEY

This early interest in biology was also fostered at school, where he was blessed with a like-minded teacher.

RS: I had a wonderful science teacher at school, a very good biology teacher, and we did marvellous experiments in school. I was doing research on fruit-fly genetics when I was sixteen, and breeding fruit flies. We had an axolotl, an embryonic salamander, that he turned into a salamander by adding thyroxine to its water. It underwent this transformation, even though its ancestors hadn't turned into salamanders for millions of years. There were things like this going on in the school science lab that I was really fascinated by.

I was so keen that, after I left school at seventeen, I got a job in a science research lab. I had a scholarship to Cambridge and some time off, nine months, and I really wanted to get on with learning about research science. That, however, was an enormous shock because I got a job in a drug company in London. I said I wanted to do biology, so I ended up as a junior technician in their main vivisection facility. This was a tremendous shock to me. I was doing biology because I loved animals, but every day my job was to get these cats ready for vivisection. Every evening, I had the worst jobs. All the rats and mice and guinea pigs that had survived toxicity tests had to be gassed. The survivors were all killed anyway, and it was my job to gas them and throw them into the incinerator. I was suddenly in this place where the most appalling things were happening to animals. Day-old chicks were being injected with LSD to see if they'd walk off the edge of a visual cliff. This was just regular pharmaceutical research, but I was quite unprepared for it, and it made me wonder what was going on.

When I did mention to one or two people there that I didn't feel terribly comfortable with this, they said, that's just an emotional reaction. Emotion has no place in science. Science is about objective facts. I began to feel that this extreme alienation from feelings was leaving something out, and I think that, more than anything, made me question the foundations of conventional mechanistic biology. I saw where it led. Of course, all this is useful research, it helps in producing new drugs and so on, but the attitude to the animals was that they're just machines. They have no feelings. They can't think. They're just part of a production line for producing these drugs. That wasn't how I thought about animals.

DC: Sheldrake went on to university at Cambridge. There, the slaughter wasn't as extensive as it had been in the pharmaceutical lab, but he still found that the study of life seemed always to run through death and dismemberment.

RS: The first thing we did to every animal or plant that we were studying was to kill it. The fact that I ended up as a biochemist didn't help, because biochemistry is all about molecules in test tubes, so to get the molecules in the test tubes, you've got to kill your animal or plant first and grind it up and then get the molecules out. It seemed to me that this way of pursuing the science of life was simply leading to more and more molecules and bits isolated from living things, but life itself was nowhere to be seen in this kind of analysis.

DC: Sheldrake became a specialist in plant physiology and developmental biology. He read his way into the history and philosophy of science, and he lived for a number of years in Asia, doing plant research in Malaya and India. One question in particular preoccupied him.

RS: How does a plant grow from a seed into a tree? The seed has very little structure. The tree has a lot. Where do all these stems and the trunk and the bark and the flowers and the leaves come from? How do they develop from a much less specialized structure? How does an animal grow from a fertilized egg, which has very little form, into something like you or me, with eyes and ears and spleen and kidney and liver and toes and so forth? All this form develops from less form. How does this work? This is still one of the big, unsolved problems of biology today. At the time I entered the field, in the 1960s and 1970s, the usual assumption was: it's all genetically programmed. It's just a matter of finding out more about genes and how they're switched on and off. But it soon became clear to me that that was not going to provide the whole answer because, if you look at the leaves and the flowers on the same plant, they both contain the same genes, so the genes alone can't explain things. Something else has to give rise to the form or the shape.

DC: Most attempts to account for the development of form in living things had focused on physical mechanisms. Cell differentiation must be cued by some chemical change or by some positional information, some evolving pattern within or between cells. Sheldrake himself did original research on the role of a plant hormone called auxin in cell differentiation. In the end, though, he concluded that none of the proposed mechanisms could satisfactorily account for the minute particulars of form. For that reason, he began to explore the possibility that morphogenesis, the origin of form, was a field phenomenon, like gravity or electromagnetism — an idea that had occurred to others as well.

RS: The term was first introduced in 1920 by a Russian called Alexander Gurwitsch, who was studying the growth of mushrooms. He was interested in how, when a mushroom grows, you have all these threads that grow in the soil, the hyphae, separate threads that then grow together and form the mushroom that we see that grows up through the soil. It's formed by a coming together of lots of separate threads that all organize themselves. Now, how did they do that? Gurwitsch thought it was because they came under the influence of a field, a bit like a magnetic field, that somehow told each bit where to go; there was a field shaping it, a field that was within and around the organism. It was like the magnetic field within and around a magnet. That was his idea.

DC: Gurwitsch was not alone in his speculations. A number of other biologists adopted the field idea and tried to model such fields both physically and mathematically. In Sheldrake's view, though, this work, though promising, remained stuck in what he has called a "theoretical limbo." Models were made of morphogenetic fields, but the question of whether they were something more than a way of speaking was not addressed. Sheldrake took the next step. He supposed, first of all, that they were actual fields, though of a type as yet unknown to physics. Then he tried to think through the implications, the first being that they must evolve.

RS: They have to change over time because species change over time, so they had to have a kind of memory. Then I hit on the idea of morphic resonance, which is the idea that each species has a kind of collective memory. Every similar thing influences all subsequent similar things over space and time. It really, when generalized, leads to the idea that the so-called laws of nature are more like habits.

DC: Morphogenetic fields are a collective memory. A beech tree becomes a beech tree by tuning in to past beech trees through the process that Sheldrake calls morphic resonance. As beech trees evolve, their morphic field evolves with them. They influence the field, as well as being influenced by it, and it is this interplay that leads him to speak of the regularities of nature in terms of habits rather than laws. A habit, however settled, remains temporary. In time, it might change. The metaphor of law, which has so far dominated the history of science, has a very different implication.

RS: In the seventeenth century, the founding fathers of modern science thought that they were uncovering the mathematical laws of nature, which were ideas in the mind of God. They didn't think these were just human models of nature. They thought they were uncovering the basic truth of nature, which was mathematical, eternal and divine. God, being omnipotent, was conceived as the cosmic emperor and also the cosmic law-enforcement agency, and this produced a theological model of laws of nature, which was very persuasive in the seventeenth century.

By the end of the eighteenth century, a lot of scientists had become deists or atheists. They dropped the idea of a god that really ran the universe, and the atheists then got rid of God altogether, but they were left with the ghost of the god of the world machine in the form of the laws of nature and a kind of universal law-enforcement-agency system, which no longer had any basis. It's a terribly inappropriate metaphor.

With the Big Bang theory in the 1960s came the idea that the universe began very small and very hot, less than the size of the head of a pin and having a very, very high temperature, with no form and no structure. Then it expanded, and everything we know has come into being — atoms, molecules, crystals, planets, galaxies, and so forth — they've all come into being in time. Everything is evolving now. The whole universe is evolving, not just life. So what about the laws of nature? Were all the laws of nature there at the moment of the Big Bang? Were they there before the Big Bang, if there was a "before the Big Bang"? It's clearly a metaphysical idea, and it's very anthropocentric as well because human laws are found only in human, civilized societies. Moreover, human laws change, they evolve, so why shouldn't the laws of nature evolve in an evolving universe? Or why shouldn't we go beyond this inappropriate metaphor? My suggestion is to adopt the alternative metaphor of habits.

DC: Sheldrake is not the first to suggest that nature may be governed by habits. The English novelist Samuel Butler had suggested as much in the nineteenth century, as did American philosopher C.S. Peirce early in the twentieth century. However, Sheldrake has gone further. He has proposed a means by which such habits might establish themselves. As an example, he cites the way in which crystals are formed.

RS: Here you have something that's part of crystallography or chemistry and supposed to have a physical explanation. If you make a new compound for the very first time, one that's never existed before, the standard view would say that the way it crystallizes is fixed in advance by the laws of electromagnetism, quantum theory, thermodynamics, and so forth. Therefore, it should crystallize the same way the first time or the billionth time. There's no difference. The laws don't change. They're the same everywhere, in all places, at all times.

However, if there's a habit principle in nature, the first time you make this crystal, a long time may pass before you get it. That chemical has no habit of crystallizing. It's a new chemical. The next time you crystallize it, there could be an influence from the first crystals. The billionth time you crystallize it, there'd be influences from all billion crystals before that, and, therefore, this accumulating influence or memory would give it a stronger tendency to crystallize. It would be getting into a groove of habit. This would predict that newly formed chemicals might be very hard to crystallize at first, and then get easier to crystallize as time goes on. That's actually what chemists find: new chemicals do get easier to crystallize as time goes on. I would say that what we're seeing here is the habit building up. What chemists say is that, oh, well, there's no mystery here. It's just because fragments of previous crystals get carried from lab to lab and act as nuclei or seeds for the crystallization process. They like telling stories about migrating chemists with beards who are supposed to carry these crystals from lab to lab, or they're supposed to be wafted around the world in the atmosphere as invisible dust particles. Here's a phenomenon that agrees with the habit approach but which chemists explain anecdotally, and this is one area where this theory could actually be tested.

DC: Lacking a chemistry laboratory, Sheldrake has not been able to carry out such a test. Nor has he yet been able to persuade anyone who does have one to take the risk of challenging the bearded chemist theory. Sheldrake has undertaken other tests of his theory, as you'll hear, but first I asked him to say a little more about the makeup of morphogenetic fields. What, in his opinion, are they? He began his answer with some remarks about fields in general.

RS: Fields are integrative. They link everything together. The gravitational field of the universe includes everything in the universe and links it all together. Secondly, as well as being integrative, fields are holistic. They're wholes. You can't have a part of a field. If you take a magnet, for example, which has a magnetic field, and you cut the north pole off the magnet, you don't get an isolated north pole; you get two magnets, each with a north and a south pole. The field, as it were, instantly regenerates to give you a complete whole.

Now, when the idea of morphogenetic fields was put forward in the 1920s, it was this that impressed the people who proposed it. They thought of the field as within and around the organism, shaping it, having a kind of pattern in space, integrating the development of the organism, and also having this holistic property of the whole being implicit in each part. If you cut a willow tree up into little bits, each little bit, each cutting, can grow into a new tree. It's a bit like cutting a magnet into bits. Each is a complete magnet. These, I think, are properties of morphogenetic fields.

What are fields made of? That's a big problem, even for the fields science recognizes. People used to think electromagnetic fields were made of the ether, the electromagnetic ether. Then in 1905 Einstein said, you don't need the ether, it doesn't exist. And what of the gravitational field? If you ask, is this in space-time?, the answer of Einstein is, no, it's not *in* space-time, it *is* space-time. The framework of space and time is a field. Now, what's that made of? These ordinary fields don't have a common-sense explanation. They're not made of subtle matter. That's the ether idea that was abandoned long ago. Instead, science is telling us that matter is made of fields. Matter is made of energy bound within fields. A proton or an electron is a vibration of energy in a proton field or an electron field. These are the fields of quantum field theory. The nature of fields is surprisingly elusive, even the well-known fields of physics.

I think morphogenetic fields are a different kind of field that's involved in shaping living organisms — the growth of embryos and plants, the inheritance of instincts in animals, the nature of memory in human beings, the behaviour of flocks of birds and schools of fish — and this also leads to the prediction of a whole range of new memory-like phenomena in nature. It's an exciting hypothesis because it takes this field idea, develops it further, and could shed light on a huge range of natural phenomena.

DC: Sheldrake's theory has had a mixed reception. His books have reached a wide audience, he's had productive dialogues with philosophically minded physicists, but he hasn't had much of a hearing so far in the citadels of scientific orthodoxy. John Maddox set the tone with his editorial in *Nature* in 1981: "Sheldrake's argument is in no sense a scientific argument," Maddox wrote. It's an "exercise in pseudo-science." Yet Sheldrake insists that his theory is a scientific theory; that is to say, it makes testable predictions, and it accounts for previously unexplained phenomena. An example is the results obtained in the rat learning trials that used to be a staple of behavioural psychology.

RS: If you train rats to learn a new trick, the more rats that learn it, the easier it should get for rats of that breed to learn the same trick. This is a prediction of the theory, which says that there's a morphic resonance from rat to rat and new habits build up. This should be transmitted by morphic resonance. Well, I found that in fact people have trained rats to learn new tricks in laboratories, so I dusted off results that were already there, the most interesting being a series of experiments that started at Harvard years ago. They started training rats to escape from water mazes. The rats, on average, made over 250 errors before they learned how to do it. Then the experimenters bred those rats, and their children made fewer errors, and their children made fewer, until they were down to thirty or forty errors — a huge improvement in learning rate. After the experiments had been going on for a while, it began to look as if this improvement involved the inheritance of acquired characteristics, which is a heresy in biology. You're not supposed to be able to inherit things that parents have learned. Animals aren't, and people aren't, except through cultural transmission — not genetically. So the experiments were challenged. The experimenters were accused of selecting only the brightest rats in the succeeding generations. Because they learned quicker, the experimenters bred from them sooner, and, in that way, there was a subtle selection going on. The person doing this, McDougall, and his assistants at Harvard changed the design. They said, okay, we'll only breed from the most stupid rats in every generation. They should, therefore, get slower and slower, according to conventional biology. But even with the most stupid rats as the parents of the next generation, they got quicker and quicker.

Then some people in Edinburgh started doing this experiment. The rats there started much quicker than the Harvard ones had originally. They took up where the Harvard ones had left off. They got quicker and quicker. They replicated the effect.

Then this was done again in Australia. The rats there got quicker and quicker in subsequent generations. What they found was that rats of another line, a breeding line that had never been trained, got quicker and quicker, too. In each generation, they'd take rats from parents that had never been exposed to this water maze, that had never had to do this task, and they also got quicker. All the rats of the breed were getting better at doing this. It was nothing to do with genes being passed on. People said, oh, well, there you are: it's refuted the Lamarckian inheritance theory; the experiment's a success, and they forgot about it. But this very long series of experiments, one of the longest in the history of rat behaviour research — and a lot of psychologists worked with rats for years — this long series of experiments produced a most intriguing result, which no one explained, which fits exactly with the predictions of morphic resonance. That's an example, I think, that shows the kind of effect one would expect.

DC: William McDougall's rat trials at Harvard and their international sequels are just one of many evidences that Sheldrake has put forward on behalf of his theory of morphic resonance. He has devised many experimental tests of the idea that living things can learn more easily what others of their kind have already learned. For those, you can refer to his published work or his website. He has suggested that his theory offers a way out of various scientific standoffs, as between nature and nurture or whether acquired characteristics can be passed on. He's argued that morphic resonance is superior to materialistic theories in explaining how memory is stored. And he's pointed to the compatibility of his hypothesis with many of the ideas of contemporary quantum physics.

One particularly rich source of confirmations has been animal behaviour, as with McDougall's rats. Anyone who has wondered at a flock of birds taking off and turning in perfect unison will easily see why. Animals often seem linked in ways that are hard to account for. One striking instance occurred in Britain in the days when home delivery of milk was still common.

RS: In the 1920s, in Southhampton, people noticed that the cream had been disappearing from the top of their milk bottles every morning, and when they watched, they found that blue tits — small birds, I think you call them chickadees in North America — were pulling the cardboard caps off the bottles and drinking the cream out of the top of these bottles. Then this turned up somewhere else, miles away. Luckily, a whole network of amateur birdwatchers started recording this all over Britain. There was a coordinated effort to record this phenomenon. Blue tits are home-loving birds, and they don't normally move more than a few miles from their home, so when this behaviour turned up more than twenty miles from where it had been seen before, it was ranked as an independent discovery. Now, the interesting thing here is that some scientists said, well, blue tits are used to pulling bark off trees and looking for things, and this was just an extension of their normal behavioural patterns. That's true, but why did it start happening with milk bottles? If it was just random, independent discoveries, you'd expect there to be a constant rate of independent discovery and then local spread, but the behaviour actually spread at an accelerating rate. It seemed that more and more blue tits were discovering this all over Britain until, by the time of World War II, they were doing it every-where. The rate of spread of the habit was measured. It seemed to accelerate. Some biologists, including Sir Alistair Hardy, who was professor of zoology at Oxford, were so impressed by this acceleration that they thought that something like telepathy must be going on, that it wasn't just normal diffusion of a habit, that the rate of change suggested that something more than that was happening.

In the Netherlands, the same thing started happening. Interestingly, when the Germans occupied the Netherlands, milk delivery stopped. Blue tits live only three or four years, so after the war, when milk deliveries began again, there wouldn't have been any surviving blue tits that remembered the golden age of free cream before the war. Yet the interesting thing there was that this habit re-established itself almost immediately all over the Netherlands when milk delivery started again. Somehow the whole blue tit population was much more prone to do this than it had been the first time around.

DC: Cream theft by blue tits is one of a number of examples Sheldrake has developed in his work about new habits spreading in animal populations at a rate impossible to explain by direct transmission. His explanation, of course, lies in his field theory of the mind. Our consciousness, he says, extends far beyond the insides of our heads, just as it does for blue tits. He has not shied away from the more radical implications of this theory, like the possibility of quasi-telepathic connections between minds. Indeed, he has tried to test it and reported his results in a book called *The Sense of Being Stared At*.

RS: The normal view within institutional science and medicine is that the mind is the brain. Mental activity is brain activity. It's all inside your head. That means, in trying to understand vision, people would say, well, when you see things, light comes into the eyes, inverted images appear on the retina, changes take place in the optic nerves and in different parts of the brain, and then, somehow, you see things inside your head. When I look around me, when I look out of the window, the trees that I'm seeing are basically patterns inside my brain.

I don't think it's like that at all. I think vision involves light coming in and changes in the brain, but then the world we're experiencing is projected out, so my image of those trees is actually outside, through a window, right where those trees are. They're in my mind, but not in my brain. My mind extends beyond my brain through fields. It's what's believed all over the world. It's what young children under the age of ten spontaneously believe, although grownup people in the West are supposed to deny this because they've been told that it's all in the brain. It doesn't feel as if it's all in the brain. It doesn't seem as if it's all in the brain. But somebody says, it's all in the brain. That's what science says. It's just an assumption. It goes completely against our experience.

Interestingly enough, a lot of philosophers of mind are now challenging this. There's a new movement called Radical Externalism, which is all the rage in the contemporary philosophy of mind — it's heavily debated at the moment. Philosophers are saying, well, when I see a tree out there, why shouldn't my image of the tree be exactly where it seems to be? How can you prove it's inside the brain? Consciousness is not the same as nerve cells. It's something different, and there's no way science at the moment can explain it at all. Now even philosophers are taking up this more traditional view. They see it as a question of philosophy, and so did Plato. I see it as science, and, therefore, I want to test it. If, when I see things, I project out visual images, and these images, as it were, touch what I'm looking at, then it means I should be able to affect things by looking at them. If I look at another person from behind when he doesn't know I'm there, I should be able to affect him by looking at him, so he might feel I'm looking at him. That person might have a feeling of being stared at — hence my book, *The Sense of Being Stared At.* I've now done lots and lots of experiments, and so have other people, that suggest that this effect is indeed real. We can tell when someone is looking at us from behind in the absence of any known kind of sensory clue, and I think that's because our vision isn't confined to the inside of the head. It involves this outward projection.

DC: When you test your theory, what kind of results do you get?

RS: The simplest tests involve people working in pairs. One sits behind the other. The subject is blindfolded. The other person either looks at them or looks away and thinks of something else, and there is a signal for the beginning of the trial, either a click or a beep. In each trial, subjects have to guess within a few seconds if they're being looked at or not. By chance, people would be right fifty percent of the time. We've done hundreds of thousands of these trials now. The average hit rate is fifty-five percent, not a lot above chance, but with those numbers, highly significant. Some people are much better than that. This is just an average. Some people are insensitive. Most people score a bit above chance. These experiments have now been widely replicated. They've been done through windows, through one-way mirrors, even through closed-circuit television. It seems people can tell when they're being looked at. There is now a considerable body of evidence that suggests this is real, and surveys show about ninety per cent of the population have had this experience, anyway, so this is a very common experience. Animals also seem to detect our gaze. My own feeling is this is an evolutionary sensitivity, that in predator-prey relations, a prey animal that could detect when a predator was looking at it would tend to survive better. I think that the evidence suggests it's real. It fits with common experience. Animals have it, too. There are good evolutionary reasons for it. And it fits with this field theory of the mind and an understanding of vision that accords with our actual experience, in which our visual world is outside us, not inside us.

DC: Sheldrake's work is often described as "controversial," but a lot of the things that he has investigated are, in fact, quite common beliefs and shared, I would suspect, by many scientists when they're away from their offices. Lots of people feel they know when they're being looked at, believe their dogs are telepathic, experience premonitions, doubt that the world is all in their heads. Lived experience and "science says" jostle uneasily in many modern minds. It's not Sheldrake's all-too-human concerns, then, that are controversial. It's his having dared to try to bring them into the world of institutional science — a daring that has, in many ways, shaped his career.

RS: It's been a long and interesting path. I've always found, even before I published any of these ideas, that there were some people who were very, very interested in thinking about the deeper questions, how things really work. There are other people who think they know the truth already, who are impatient with any speculation that goes beyond the limits of what they think is permissible. There are very strong taboos against certain kinds of speculation. These taboos, I think, are related to a rigid form of dogmatic rationalism: the mind is in the brain, there's nothing in the world that science can't explain, and everything is explicable in terms of regular physics and chemistry. That view is often associated with a powerful taboo against telepathy and other so-called psychic phenomena.

Now, people who are true believers of this kind — and I've come to recognize this as a kind of scientific fundamentalism — these scientific fundamentalists simply aren't interested in discussing any of these bigger questions. They know the truth. All these other ideas are heretical, and they should be suppressed or denied or ridiculed.

Right from the beginning, I've had a mixture of responses. One has been open-minded curiosity from people who have said, okay, that's an interesting idea, but where's the evidence? I've said, here's some evidence. And they've said, couldn't there be something wrong with it? Reasonable discussion, reasonable skepticism. I'm used to reasonable skepticism. I consider myself a skeptic as well. I think skepticism involves challenging established dogmas, not just defending them. When my first book was published in 1981 — that's a long time ago — there were a lot of people who were very interested by it and discussed it and were excited. Here's a new idea. It has a lot of implications. Here in Britain, the *New Scientist* magazine ran a big story about it. I wrote an article for *New Scientist*. There were leaders and editorials in papers like *The Guardian*. There was serious, intelligent discussion on the BBC and in the media until, after three or four months, there came this astonishing attack on the whole thing — the famous editorial in *Nature* written by its editor, Sir John Maddox, called "A Book for Burning." It was an extremely violent attack on my book, saying it was absolutely infuriating, it was the best candidate for burning there'd been for many years, and so on. Basically, what Maddox was doing was trying to excommunicate me from the world of science. In fact, only about ten years ago, when he was interviewed about this famous editorial, he actually said on British television, Sheldrake deserved to be condemned for exactly the same reasons the Pope condemned Galileo. It's heresy. I think it's more revealing, really, that he was comparing himself to the Pope rather than me to Galileo, but he felt that he needed to do something about this and stop this rot spreading through science.

After that, this heretic label which he tried to attach to me has had quite an enduring influence. It means that it would be hard to get a mainstream grant for the kind of research I do or anything like that. Ever since then, my work's been controversial, and the work I've done on telepathy and on the sense of being stared at has drawn the attention of these organized groups of skeptics. They're scientific vigilantes, except most of them aren't even scientists, they're scientific fundamentalists, and they'll do anything they can to counteract what I'm doing by trying to mock and ridicule it, using techniques that are quite often unfair and very unscientific. I'm used to that kind of opposition, but what strikes me over and over again is that, within the scientific world, these scientific fundamentalists are a minority. A lot of journalists and people in the media assume that they speak for the scientific community. They don't. They speak for themselves and for these rather extremist advocacy organizations. Within the scientific community, I discover a lot of open-mindedness, in fact, more now than ever before, because the old certainties are crumbling. People used to think: just sequence the human genome and we'll understand the secret of human life. Well, it's been sequenced, and it's rather a damp squib. We still don't understand most of the basic problems. Then, people used to think in the 1990s, the decade of the brain, so-called: scan everything in the brain, work out what all these different tissues are doing, and we'll understand the mind. The answer is, we don't. The problems are as great as ever — greater, in fact, because the more we understand about the brain, the less we can have fantasies that it will explain the mind. We now know what's actually happening in the brain, and it's so different from our conscious experience of the world that the gap has become huger than before. It's in sharper focus than before.

Within biology, I think, there's a growing sense that "business as usual" isn't going to go on working for much longer. This narrow, dogmatic refusal to discuss anything beyond the most conventional forms of materialism is now actually a minority position defended by these vigilante groups. There are an awful lot of people in science and in medicine who are thinking more widely. At the moment, however, they're rather shy to come out in public and say so. It's a bit like gays in the 1950s. Science is full of people who are thinking heretical thoughts, but they don't dare say so to their colleagues for fear of being thought ridiculous or stupid or heretical or something.

DC: This language of heresy and fundamentalism and dogma — it's quite clear where it comes from. Why is that the appropriate language in which to discuss science? Is there a sense in which science has been a church?

RS: I think that it's actually been a kind of church for a long time. Its self-image and its mythology picture science as the struggle for freedom of thought against the oppression of the church, with the story of Galileo's persecution by the Pope and the cardinals being used over and over again. Thus, scientists are heroic, freethinking people, struggling against the dogma of the church. In fact, the church, at least in countries like Britain, has very little power nowadays, and science has a huge amount and has become a kind of dogmatic system of its own.

I think we have to recognize that this was implicit in the very first vision of scientific institutions by Sir Francis Bacon at the beginning of the seventeenth century. He was the first person who had a political vision of science, and he was a politician. He was Lord Chancellor of England. He was used to running the government. He wrote a book,

New Atlantis, which was a kind of utopian, visionary book, saying that, through organized science and technology — he didn't use those words, those are the words we now use — man can achieve dominion over nature, conquer nature and use nature for his own ends. This vision of science as something with which to dominate nature for human profit and gain was built into science in the early seventeenth century. How this should be done, according to Bacon, was to have a kind of central, government-funded college. The people who composed this college would wear long robes — they would be a kind of scientific priesthood — and it would be state sponsored, so there was no separation of science and state. Right from the beginning, there's been a linkage of science and state. That led to the Royal Society, which was founded soon after the Restoration of the monarchy in England in 1660, and, in other countries, to the founding of national academies of sciences. These are all directly descended from Bacon's vision of a kind of scientific priesthood which would flourish under state patronage and be an essential part of the government's way of having power and control, not just over people, but over nature. I think this religious imagery, this religious image of science as a kind of church is actually very deeply built into the identity of institutional science, and it accounts for a lot of the very weird features of science that puzzle people who aren't in the scientific world: how intolerant scientists can be, how authoritarian, how top-down and undemocratic the structure of science is. This is like the unreformed church.

The other thing is that science has got away with the idea that, whereas there are many kinds of Christian sects, many kinds of religion, there's just one kind of science. In each country, there's one central institution that tells you what the right kind is, and, moreover, there's only one kind of science in every country in the world. No system of thought has ever conquered the whole world as science has, and I think it has built into it a kind of monolithic or authoritarian structure that's got worse and worse since World War II. In the nineteenth century, a lot of science was done independently by independent researchers. Charles Darwin was an example. Darwin never had a government grant. He didn't have an academic post. He lived as a private gentleman. He was quite well off. He did research that interested him. He never had to get peer review approval, ethics committee approval, and so forth. He did what he wanted. A lot of science was done that way. Since World War II, it's become more and more centralized, more and more institutionalized, and there's almost no independent science going on at the moment. It's all under the control of government funding agencies or corporate funding agencies, and these are all working on an anonymous peer review system, so people's work is usually evaluated by their rivals, anonymously, and based on short-term grants. People are terribly afraid of stepping out of line for fear that they won't get their grant renewed. Most scientists are on a very short leash.

DC: The domination of science by state and corporate agendas is well known to the general public, in Sheldrake's opinion, and is, he thinks, a prime reason for public cynicism. His proposal, which he has made publicly, is that one percent of state science funding should be apportioned by popular decision. Let popular organizations, like trade unions, horticultural societies, or environmental groups direct the funds to the research that they would like to see done. Science, he says, is due for a reformation, something not unlike the shattering of the monolithic Christianity of the late Middle Ages. One reform that he would particularly like to see is a renewal of popular participation in scientific research.

RS: I think one of the changes we need in science is a much more participatory attitude, not only public participation in science funding decisions but also more public participation in the actual doing of science. In the nineteenth century, a lot of science was done by amateurs. Such people still exist — the gardeners and naturalists are still there — but they've mostly been squeezed out of institutional science and marginalized. They often know an awful lot more about living species, about animal behaviour, about plants than professionals working within science, who are busy going to meetings and filling in grant applications. Many people have been excluded from the scientific endeavour who, I think, would greatly enrich it if they were brought back in. A lot of my own research is based on involving non-professional scientists in research. Many of my own projects are happening in schools and colleges at the moment. On my website, I have a number of experiments that anyone can do just by logging on and doing on-line tests, and already thousands of people have taken part. I think this is one way forward: making science much more participatory, making it much less of an exclusive priesthood, which I think has just alienated many people from science and put off young people from going into it.

DC: Popular participation is one way in which Sheldrake would like to see science renewed. Another is to release it from the dead hand of mechanism, reductionism, and the outworn metaphor of scientific laws. In his view, contemporary scientific findings have given us many reasons to return to the once-universal belief that nature is alive, but much of science remains trapped in an obsolete philosophy.

RS: We normally have the idea of nature as governed by fixed laws, and the usual assumption in science is that these laws were fixed at the moment of the Big Bang. As my friend Terence McKenna used to say, "Modern science is based on the principle, Give us one free miracle, and we'll explain the rest." The one free miracle is the origin of the entire universe and all the laws that govern it, from nothing, in a single instant; thereafter, it's supposed to be business as usual. This idea that all the laws of nature are completely fixed and all the constants are fixed is simply assumed by most scientists. I think it has no basis other than an outmoded metaphysics. It's a habit of thought.

This leads to a whole range of other problems. Then you say, well, if they're fixed exactly right for us to exist, then they must either have been fixed by some kind of deistic god, a sort of supreme intelligence, who fine-tuned all the laws and constants, pressed the "start" button of the universe, and then stepped back and was never heard of again. Or, alternatively, there's the view favoured by many modern cosmologists that there are billions of actual universes, of which ours just happens to be one, and the laws and constants in our universe occur just by chance. According to them, all these other universes actually exist. Lord Rees, the president of the Royal Society, is a firm believer in that principle. It's extraordinary to me that something like telepathy can be controversial when the postulation of billions of totally unobserved universes can pass as reasonable mainstream science. It's a remarkable thing. Anyway, all that kind of speculation is unnecessary with morphic resonance. If the regularities of nature are like habits, evolution, at a cosmic and biological level, is an interplay of habit and creativity. All sorts of new forms, new patterns, new possibilities are coming up all the time. Most of them are not viable. They don't survive. The successful ones are repeated and become new habits. Everything hangs together in nature because of natural selection. Everything's adapted to everything else because it's had to evolve. We're returning to a view of an organic nature, a living nature, which is what people believed in before the seventeenth-century revolution in science. Then everyone thought nature was alive and organic, but now the difference is that we see it as evolutionary, as developmental, as having a kind of history. They thought it was cyclical or eternal. It went in cycles; it didn't really evolve, and if anything, it just got worse. I think the new vision of the universe as a developing organism is very exciting, and habits are an integral part of this vision of a living nature.

Rationality and Ritual

BRIAN WYNNE

It's believed because it's true. That was the explanation of scientific truth. But when you actually look at why scientists believe the things they believe, it's rubbish. They don't believe them because they're true. They might be true, but that's not why they believe them.

— Brian Wynne

Technological science exerts a pervasive influence on contemporary life. It determines much of what we do and almost all of how we do it. Yet science and technology lie almost completely outside the realm of political decision. No electorate ever voted to split atoms or insert DNA from one organism into another. No legislature ever authorized the iPod or the Internet. Our civilization, consequently, is caught in a profound paradox. We glorify freedom and choice, but we submit to the transformation of our culture by techno-science as to a virtual fate.

The politics of scientific knowledge is a subject that has long pre-occupied Brian Wynne, a professor at Lancaster University in the north of England and the associate director of an institute that studies the social and economic aspects of genetic technologies. In the 1970s, he joined the influential Science Studies Unit at the University of Edinburgh and became part of an academic movement to radically revise received wisdom in the sociology of scientific knowledge. At the time, science was frequently characterized in the terms that had been put forward by American sociologist Robert Merton in an influential essay called "The Normative Structure of Science." Science, Merton said, was guided by four principles: communalism, universalism, disinterestedness, and organized skepticism. These became widely known by their acronym CUDOS. The gist of CUDOS is that science has one overriding commitment — the truth — and in pursuit of this grail, scientists are expected to sacrifice all parochial concerns — that's the universalism — all private interests — the communalism and disinterestedness — and all prior intellectual commitments — the organized skepticism. Brian Wynne and his colleagues challenged this heroic image of science. They argued that science was an inherently social enterprise and just as capable of parochialism, self-interest, and superstition as any other social institution. For them, the genius of science lay in its social organization and not in some heroic ability to stand outside society.

One of the ways in which this new sociology of scientific knowledge was pursued was through studies of what actually goes on in scientific laboratories. Another was through the analysis of scientific controversies, past and present, in which "the facts of the matter" were in dispute and one could therefore observe scientific knowledge still, as it were, under construction. Brian Wynne took a different route. He wanted to understand how scientific knowledge exerts its authority in public arenas. In our interview, he told me about the epiphany that had started him on this road. He grew up in a village in the northwest of England, and his academic ability got him to Cambridge, where, in 1971, he completed a PhD in material science, the branch of physics that studies the properties of engineering materials. Up to that point, he said, he had never really thought about the politics of science. But then he had a fateful conversation.

BRIAN WYNNE

My PhD supervisor at Cambridge said to me, well, do you want to do a post-doc? I'd had a ball, really, so I thought, why not? It was the obvious thing to do. I said, yeah, great, I'd love to. So he said, well, just write down a few thoughts about what you'd like to do, and hand them in, and we'll talk about them in a week or two. So I thought about it. That was the time when oil prices began to go through the roof, and energy, energy, energy was on everybody's mind. I just thought, well, as a material scientist, I ought to be able to do something helpful in all of this. What about smart materials for energy efficiency and energy saving? I wrote down a few ideas around that kind of thing and went to see him, and he treated me as if I'd just come from planet Mars, you know. He was very dismissive. I'd got on with him fine as a supervisor, but he was very dismissive, and I was really disappointed. I went off, and I was having a drink with a friend in the Cavendish lab, a physicist. I was lamenting to Pete about what had gone on because I was so puzzled and disappointed, and he said, Brian, just stop and think about it. Where does the funding money in your department come from? We're speaking of the Cambridge Material Science Department, one of the leading material science departments in the world. He just asked me that very simple question, and I didn't know the answer. I'd been there for six years, and I didn't know the answer to that simple question. This friend was much more politically alive and aware than I was.

I then stopped and looked and did a bit of prodding around to see. Of course, not surprisingly, most of the bloody money that was keeping the department afloat was coming from military or quasi-military outfits. If I'd wanted to do a post-doc on something like the next generation of alloys for tanks or missiles or God knows what, there'd have been money flowing like there was no tomorrow, but for something that I thought was scientifically interesting and socially useful, there was no interest, and presumably, therefore, no money.

At that point I started to think, well, maybe this isn't what I want to do for the rest of my life. I really don't want to give my life to being an appendage of the military industrial complex.

DAVID CAYLEY

Brian Wynne's revelation pushed him from science into the emerging field of science studies. He went to the University of Edinburgh and immersed himself in the literature that was beginning to put science in a new light. One of the crucial texts for almost everyone involved in the field was written by the American philosopher Thomas Kuhn.

BW: Kuhn's famous book of 1962, *The Structure of Scientific Revolutions*, really delineated the ways in which science is a cultural practice. It's cultural in the sense that there are forms of closure around particular, effectively dogmatic commitments to ways of seeing nature. These are reflected in methodological commitments and in theoretical paradigms, as he called them. That was really where the famous term "paradigm" came from. The particular theoretical commitments in any given speciality are not necessarily the objects of scientific skepticism and skeptical critical testing, but they are actually the framework within which observation and analysis and testing take place. There are important elements, in other words, of dogma. Kuhn wrote a famous paper called "The Functional Role of Dogma in Science." A certain ambiguity, in Kuhn's view, is essential to modern science. He called it the essential tension.

DC: This tension in science, between free inquiry and dogmatic commitment, became a key idea for Wynne. It explained, for example, why scientific theories can withstand a certain amount of contrary evidence, or what Thomas Kuhn called anomalies.

BW: It's not the case that when we get an anomaly that can't be explained by the existing theoretical framework — and this applies in whichever speciality we're talking about, whether it's solid-state physics or orbital chemistry or whatever else it might be — it is not the case that, when we get an anomaly that can't be explained, we throw away the theory and look for a better one. As Kuhn put it, every theory is born refuted. There are always anomalies. Every single theory that exists on the face of the planet has always had anomalies associated with it. The key thing is whether the scientists can persuade themselves to suspend and shelve the anomalies for the time being in the faith that eventually they'll be explained. As we develop the theory we will be able to explain those.

DC: Kuhn helped Wynne to see science as a cultural enterprise — a mixture, Wynne would say in his first book, of "rationality and ritual." This was surprising, even shocking, because natural science until the 1970s had enjoyed an extraordinary privilege. Other forms of knowledge might be shaped by circumstances, but science was a transparent revelation of nature itself. What scientists believed, the story went, was simply what was the case.

BW: It's believed because it's true. That was the explanation of scientific truth. But when you actually look at why scientists believe the things they believe, it's rubbish. They don't believe them because they're true. They might be true, but that's not why they believe them. Having been through six years of scientific education, I can speak from personal experience. I know that's the case. Then a key question is, okay, so if we're saying that scientific truth requires a sociological explanation, does that mean that scientific truth is just *socially constructed*, as people often say? Does that mean nature has no voice in the matter? That's where I think people often make a crucial mistake. There's no logical reason at all why saying that something is socially negotiated and constructed as a belief should imply that nature's not playing a role in it. Most of the time when we're constructing and negotiating — sociologically negotiating — what counts as valid knowledge, we're also building in the question, does it work? You know, we don't particularly want knowledge that doesn't work. What would be the point of that? It's a perfectly social thing to do, to say, let's make sure this knowledge works. In other words, we want it to be as true as possible. However, when you go further into that question, you say, well, okay, works for what purpose? What counts as working? That's a very material and social question. People have got different views about what counts as working, because they have different views of the proper purposes of this knowledge, and that's where we get into a lot of the contemporary issues.

DC: Wynne's interest in the proper purpose of scientific knowledge pushed him in a political direction. He shared his Edinburgh colleagues' interest in the question of how scientific knowledge is made and institutionalized — what he calls the esoteric side of the new sociology

— but what fascinated him most was the face that science turned towards the public.

BW: When I came from Cambridge to Edinburgh, I was really just waking up to the sheer nakedness and hugeness of the political world within which science is brought into being and funded and used and all that kind of thing. I was opening my eyes to all this esoteric stuff about how scientific knowledge is made, but I was also interested in politics. The environment was increasingly becoming an issue, and then nuclear power. I was always interested in scientific knowledge in public arenas, where it's intersecting not only with contradictory scientific knowledge, but also with public views.

DC: Wynne found inspiration for this political turn in the work of the late American sociologist Dorothy Nelkin, who then taught at Cornell. She was studying the role of science in political controversies, such as the one then surrounding the building of a nuclear power plant near Cornell on Lake Cayuga.

BW: Dot's approach was very much, well, these scientific pronouncements about whether the risks are too high or the risks are acceptable are being shaped by interests, by political interests. This lot are in the pay of the government or the aircraft industry or the nuclear industry, so they're saying this. These scientists lean towards the environment, so they're saying the other. Then these two different scientific perspectives come together in some decision-making context, and she looked at what happens next. What she found was very interesting and contrary to the usual scientific mythology, which is that when you put opposing scientific views together as if they were hypotheses, the truth comes out of the mix. You know, one side tests the other side, the other side tests the first side, and out of that mix comes the truth. There's consensus. That's the standard mythology about how scientific knowledge goes on and it how it progresses. What Dot was showing was that in public arenas, time and again, what happens when two opposing scientific perspectives confront each other — around nuclear risks, say — what actually happens is that each side just elaborates more and more. The thing gets more polarized, more technically elaborated, new arguments are brought in, but each side is entrenching itself and digging deeper rather than actually coming up with some sort of consensual and singular scientific truth. Her observation raises all sorts of questions: how are these scientific views being framed? Is it the framing, maybe, which is actually the real issue? That's a social thing.

DC: Dorothy Nelkin found that in public arenas antagonistic scientific positions tend to entrench themselves rather than moving to the middle ground. This suggested to Wynne that the content of the knowledge in question might not be as important as its framing, the master story which determines how the knowledge is deployed. He confirmed this finding for himself when he participated in a public inquiry on nuclear power in 1977. The issue was whether a nuclear fuel reprocessing plant, known as the THORP plant, should be added to the already massive British nuclear energy complex on the Cumbrian coast at Sellafield, not far from where he grew up. The Windscale Inquiry, as it was called, was conducted by a British High Court Judge, Mr. Justice Roger Parker, who presided over a hundred days of public hearings and then ruled that the building of the plant could proceed. Wynne followed the entire proceeding as both an advocate against the THORP plant and an observer. Five years later, in 1982, he published a book on the Windscale Inquiry called *Rationality and Ritual*. It argues that the inquiry was an elaborate ritual in which a political decision was disguised as a scientific judgment.

BW: In that book, I talked about and analyzed how the judge in the inquiry had used a very empiricist notion of rationality — that we go out there in nature and find the answer. Nature will tell us. All we need to do is to look at things in a disciplined way using the scientific method, and then we'll discover the answer that nature tells. Is it safe or not? Is the THORP plant at Sellafield safe or not? What I was trying to tell him was, look, the THORP plant is not in being yet. It doesn't exist. It's not had any impact on nature. We can't go out and look at it. All we can do, actually, is look at the prospect and also at the retrospect. What did experts say about the behaviour of previous plants from which we now have evidence we can look at? Did they fulfill the promises? I said to him that when we actually look at the evidence, we find, well, no, they didn't. So what do we make of the promises they're making about this plant, which is twenty times more radioactively intense than its predecessors in terms of the materials it's using, et cetera? It's never been done before, anywhere in the world. Why don't you take seriously the fact that they got their promises wrong last time? Maybe they'll get their promises wrong this time, too.

Instead of that, he just said, no, go out and measure the environmental consequences. Is it too risky or not? I was saying, well, you know, whether or not the THORP plant should be built is, in the end, a political issue. It's legitimate either way. You make that choice according to various wider senses of society — what kind of society you want, what kind of society you want to avoid, and so on and so forth. That's a process which, of course, involves evidence and reason and science, but it involves them in partial chunks. Science doesn't actually encompass the whole thing. Parker presided at the inquiry and presented its report as if science had dictated the truth and lawyers had helped to provide the discipline by which this truth was revealed. But to present things this way is to present a myth of how we actually go about making these decisions. It's using rationality as a ritual form of authority. There's a sense in which rationality is a kind of distraction in that public forum. I think you can see the same thing at work in many of the ways in which risk discourse is used in modern society. It's not just a question of nuclear power but also of biotechnology, GM crops, and a variety of other things. I want to say, let's do risk assessment, yes, but let's also unpack the social and political commitments that are actually being enacted through that apparently purely scientific discourse.

DC: The Windscale Inquiry was a political initiation for Brian Wynne. He told the BBC at the time that it was "like being sucked into a whirlpool." It had this character for him, in part, because of the way the inquiry was conducted. A manifestly political decision, he believed, was being camouflaged as a technical and scientific decision. The occasion was also fraught for him because of what was at stake. Windscale, later known as Sellafield, dominates a stretch of the northwest coast of England not far from where Brian Wynne grew up. It began as a military installation, producing munitions during World War II and materials for nuclear weapons after 1947. In the early 1950s, nuclear reactors for civil use were also added, and the site gradually grew into a complex so vast that both the Irish and the Norwegian governments eventually asked that it be closed. The Windscale complex also threw a long a shadow on the surrounding English countryside, particularly because of an incident that took place in 1957.

BW: One of the two military production piles caught fire and burned out of control for several days. The people in charge just clammed up. Completely secret. No announcements. Nothing. Stuff was blowing out over local farms and, in fact, right over Patterdale, where I was living at the time — I was, what, ten years old in 1957. It burned for several days and emitted a lot of radioactivity, as it turned out, over the country and beyond. They didn't know how to control it, fundamentally. They didn't have a plan B for that eventuality. They decided that the only thing they could do was to take the risk of quenching the fire with water, but they were scared silly that doing this might actually cause hydrogen to be generated, and then there would be a hydrogen explosion, in which case it would have just blown up like Chernobyl. They basically just fire-hosed water into the core — at that point red hot — to see if they could quench it. It was really like fingers-crossed time, you know. They knew that they were going to die, no question, if the situation went the wrong way. It worked. They managed to quench the fire, and people on the inside say that, once they'd actually got it quenched, they essentially just locked the door and threw the key away. They didn't want to know about the molten mess inside. They've since been busy decommissioning that site, and they've taken down the original pile. It's in the process of being cleaned up now.

DC: The 1957 Windscale fire and the secrecy surrounding it was important background for the 1977 Windscale Inquiry. It was also well within living memory when radioactive isotopes again began falling on the Cumbrian countryside, this time in 1986, after the nuclear reactor at Chernobyl blew up. Wynne studied the unfortunate consequences for the local sheep farmers, and his research gives decisive support to his earlier statements about the role of dogma in science. The tale begins with the radioactive cloud that formed over Chernobyl and then dispersed over Northern and Western Europe.

BW: What happened, first of all, was that the fallout cloud arrived over Britain, containing mainly radiocesium but also radioiodine and some strontium as well, from Chernobyl. It wafted around over Scandinavia, went down over continental Europe, and then came up again over Britain. This was probably about a week after the accident itself. It had already dumped a lot of radiocesium on Scandinavia and affected reindeer in Northern Scandinavia, too. Radiocesium is very prone to rainfall. In other words, if you get rainfall coinciding with the existence of the cloud in a particular area, a lot of radiocesium will be rained out of the cloud onto the ground. That's what happened over Cumbria. In fact, I was in the Lakes walking that day. I remember that whole weekend very well because there were very, very heavy thundershowers. We got a real dousing of rain, and I didn't know it at the time, but it was actually pretty radioactive stuff that we were getting.

DC: This radioactive rain was initially said to be harmless. Residents were told the water was safe, and the farmers were assured that there was no danger of contamination of their sheep — sheep-rearing on the bare hills, or fells, being one of the mainstays of the local economy. Even so, the region's history created a certain unease.

BW: Most of the local farming population which was affected by the Chernobyl fallout had been there in 1957, if not as adults, then as kids. That kind of farming culture is pretty stable, even these days. Farms pass from one generation to the next. Parents tell their sons and daughters loads of stuff. A lot of what passes on is sheep farming expertise, but, of course, history is also remembered, and part of that history was what happened in '57, when we got dumped on by radioactivity from the fire, and nobody told us about it. When the Chernobyl thing comes around, and the experts are saying, oh, no chance of any risks here, you know, no problem, you don't need to do anything here, one of the immediate reactions is, how do we know whether to believe these people or not? How can we trust them? Of course, the next question is, well, what's the track record, what's the previous experience? For a lot of the people there, the previous experience was what happened in the '57 fire. The authorities lied, they dissembled, they just kept quiet for several days and never told us the truth about anything. Why should we believe them this time around? It's a perfectly logical, reasonable way of arriving at a judgment based on the information they're being given.

DC: These suspicions would be aggravated by subsequent events. At first, the farmers were told that everything was all right, but then the scientists were forced to revise their story.

BW: The scientists said initially, no problem, there's no need for any worry. Even people who were collecting rainwater for drinking off the roofs of their houses were told, you can drink it — there's no problem. Your dose won't be at all significant. So everybody thought, okay, carry on, then. Within a month or so, by about mid-June, the story changed. The scientists had been monitoring, of course, for various things, and they'd monitored lambs in the Lake District and found that they were above the European action level, i.e., the level of radiation allowed by European Union regulations, which is a thousand becquerels per kilo of live weight of whatever animal you're looking at. They had to say, okay, we've got to impose a ban. But — and this was the reassuring thing to the farmers — this ban will only need to last for about three weeks. Within a month it'll be okay.

The reason they said that was because none of the levels that they'd found were more than twice as high as the action level. Every measurement they'd made was less than two thousand becquerels per kilo, and the biological half-life of radiocesium in sheep is around about twenty-one days. The biological half-life is different from the physical half-life because it's a matter of how fast the sheep excrete the stuff. That's why the scientists said that in twenty-one days all of those levels will be half what they are now. It'll all be beneath the one-thousand-becquerel level within three weeks, so if we say a month for this ban, that's going to cover it. That's how they reassured the farmers. However, when the month was up in mid-July, their monitoring was finding that the levels hadn't actually come down at all. They were still recording at above the action level, so their prediction — their second prediction by this time — was still failing and still refuted by experience.

At that point, they were forced to actually introduce an indefinite ban. That's when, as it were, the proverbial hit the fan. In the Lakes, in hill areas, you don't have that much natural pasture, and none in winter. Basically, what you have to do in the Lake District is to produce a very big crop of lambs in the spring. You then fatten them through the summer, but you have to sell that lamb crop before the winter starts because you don't have any feed for them. They'd all starve if you kept them. The normal practice is to sell them in the October sales, and they then get fattened up on lowland pastures somewhere and sold for meat. With an indefinite ban, the farmers were looking at the prospect of not actually being able to sell their sheep, and, since they wouldn't be able to feed them without the enormous cost of buying hay, they faced disaster.

DC: The wool of the contaminated sheep was marked with an orange dye. The farmers were put into a damned-if-you-do, damned-if-you-don't position. They could sell their contaminated sheep for the next-to-nothing they would fetch, or they could believe the scientists who kept predicting that the problem was about to go away.

BW: Even when they imposed the indefinite ban, scientists still stuck with their original belief that the contamination was going to come down. It was just that it was taking a bit longer. This was a classic case of dogma being reproduced. They were still stuck within their existing conviction that the contamination would come down. They couldn't believe anything else, so they continued to say to the farmers, okay, it might be taking a bit longer, and, in the meantime, we need to mark your sheep. And if you need to sell them, well, you can sell them. Of course, this advice ignored the practical reality that the farmers would have to sell the sheep for nothing. They'd be disastrously penalized financially. So the farmers held on — by now we're into August and September of 1986 — a bit longer, and a bit longer, and a bit longer, in the hope that, as scientists were telling them, the contamination would fall beneath the action level. Then they could sell their sheep as uncontaminated and unblighted by that orange mark on their fleece. A lot of the farmers held on, and then they felt like they'd been completely deceived by the scientists because the level of contamination didn't come down. They began to think — and I've got quotes on this in the articles I've written from interviews with the farmers — that the scientists were in cahoots with the government to put them out of business and that the government favoured tourism over farming. It was a complete disaster bureaucratically, economically, and, indeed, scientifically as well.

DC: The theory that the government wanted them out of business was just one of the explanations that the aggrieved farmers came up with. Another popular view was that the Chernobyl cloud was being blamed for pollution that was actually the result of the Windscale fire in 1957, and that this was what accounted for the unexpected persistence of the contamination. The real reason turned out to be less sinister but equally interesting for a sociologist of science like Brian Wynne. The scientists, it emerged, hadn't known as much as they thought about the behaviour of radiocesium in the type of soil on which the sheep were grazing.

BW: It turned out that the reason the scientists had been making this mistake was because they had been operating with a model that described the behaviour of radiocesium in alkaline clay soils, which are lowland soils. These are not the kind of soils you get in Lake District fells, which are acid, peaty, organic soils. In the acid, peaty, organic soils, once radiocesium is washed off the vegetation by further rain and gets into the soil, it remains chemically and biologically mobile. It's in the soil, but it's available to be absorbed by the root systems of plants.

DC: It's not chemically bound, as it would be in an alkaline soil.

BW: Exactly. In alkaline soils, the aluminum silicates absorb cesium, which is then immobilized. It can't be taken up into the vegetation and recontaminate the sheep. That was the assumption the scientists were making, but they didn't reflect on the assumptions they were making or ask themselves, are these assumptions valid in the Lake District fells? That's how they made their mistake. That was the basis of the false prediction that led to these false reassurances.

DC: Wynne was eventually able to track the source of the information on which the scientists in charge of the Cumbrian ban had relied. The research, it turned out, had been done approximately twenty years before by Britain's Atomic Energy Research Establishment at Harwell in Oxfordshire. Physicists there had looked at the behaviour of radiocesium in a variety of soil types, including the type found on the Cumbrian fells, but they had been interested only in the distribution and depth of penetration of the radioactive material. They had paid no attention at all to the problem that proved critical in the hills of Cumbria.

BW: They weren't looking at the issue of chemical and biological mobility and recontamination of any vegetation. Their report didn't even talk about vegetation. They were concerned with human exposure to radioactive fallout. This was their risk model, and it determined what they were looking at, and what was relevant and what wasn't relevant, and what was just neglected. They recognized that when rainfall occurs, radioactivity leeches into the soil, but their question was, how long does it take until the radioactivity gets deep enough into different kinds of soil that an average human being standing on the surface of that contaminated ground will no longer get an exposure to their gonads that might affect their reproductive capabilities? That was the risk model they were using. That was quite explicit. I thought, well, yeah, okay, fair enough, that would be one possible exposure model from weapons testing in the atmosphere, but they just didn't look at food chain models at all. They just didn't think about food contamination as a potential issue. They'd missed the trick in terms of the Chernobyl fallout and the issue about sheep contamination.

DC: It was bad enough, from the point of view of the sheep farmers, that the scientists they were obliged to trust and bound to obey were relying on research that didn't fit their case. The story got even more interesting when Wynne discovered that the relevant research actually existed but had been completely overlooked. In the 1960s, at the same time as the scientists working at Harwell were testing their purely physical model of radioactive soils, researchers at the Agricultural Research Council, the ARC, had investigated the question at issue in Cumbria. How long would radiocesium remain biologically active in contaminated soils?

BW: The ARC researchers had not used experimental plots like the Harwell scientists but had actually worked out in the environment in agricultural settings, with different types of soils and different types of vegetation. Their report stated quite unequivocally that radiocesium will remain more mobile in acid, organic, peaty soils than it will in alkaline clay soils. So there actually were scientists who were stating quite unequivocally, it'll behave differently. I was puzzled. How could it be that two competent, well-qualified scientific institutes in the same country, operating with the same issue, radioactivity and potential environmental consequences, had come to such different conclusions? I think the answer is that the physicists at Harwell, the nuclear research establishment, had all come out of nuclear engineering and nuclear physics — perhaps a bit of radiobiology, but with physics backgrounds nevertheless. The agricultural researchers at the Agricultural Research Council, on the other hand, were biologists and environmental scientists. Each was operating with different kinds of experimental habits, different kinds of theoretical resources, different models of what was going on, and different interests in terms of what was salient. The Agricultural Research Council were interested in agriculture — in vegetation, soils, weather patterns, and how these things might impact upon the environmental distribution of radioactivity. I mean, it's an interesting sociology of science story in its own right. Here were two different scientific cultures just not speaking to each other. Never the twain shall meet. They were producing different stuff and different options, if you like, in terms of what the scientists in 1986 were going to know, to be able to advise government and the farmers about the likely consequence of the Chernobyl fallout.

DC: Wynne's account of how different paradigms produce different research programs explains why the scientists at the Atomic Energy

Research Establishment were interested in different questions than the scientists at the Agricultural Research Council. But why did the scientists in Cumbria in 1986 turn to the wrong research? He investigated that question, too, and that investigation produced what may be the most interesting twist of all in this scientific detective story. The work done at the Agricultural Research Council had been absorbed into a much larger United Nations study of the effects of radioactive fallout on human populations. Once aggregated with other research, it simply sank below the horizon of the scientists in charge of the sheep bans in Cumbria.

BW: That previous work just simply disappeared. It was no longer practised knowledge, and so it was no longer known. There was not just a process of learning going on. There was a process of "ignorancing" going on as well. Nobody's deliberately doing it. Nobody's deliberately covering anything up. It's just the way things go on. One of the reasons I want to write this up is because it's got huge policy implications. It teaches us, or ought to teach us, to be a bit more modest about what we think we know. These things just happen as a result of culture and history. Practice moves on, and we actually forget things that we once knew. What do we think we know about the environmental consequences of stuff like GM crops, for example? Well, maybe we don't know that much. Maybe we don't control the risks — intellectually control them, I mean — in the sense of being capable of predicting them, as we think we do. And maybe recognizing that might induce us to be a bit more modest about such things. That's one of the major practical policy implications of this kind of historical reflection.

DC: Cumbria has experienced lasting effects from the Chernobyl fallout and was probably one of the worst affected areas outside the immediate region of the explosion. An article in *New Scientist* five years later, in 1991, reported that half a million sheep on some six hundred holdings were still under some kind of restriction. A few farms continue to be affected to this day. This was certainly not all the fault of the British scientists monitoring meat contamination, but had these scientists been aware of all that they didn't know, some of the worst consequences could have been mitigated. The sheep, for example, could have been brought down from the fells and fed on uncontaminated hay in the valleys until the worst was past. They were left on the contaminated pasture because the scientists remained dogmatically committed to an inaccurate model of what was going on. This is the enduring lesson of the whole affair for Brian Wynne — not that everybody can or should know everything, but that it's good to know what you don't know.

Wynne is currently the associate director and principal investigator with the Centre for Social and Economic Aspects of Genomics, CESAGEN. Its purpose, as the name suggests, is to study the ways in which social and economic interests are shaping the research agenda of genetics and to try to understand and assert what the public good requires in this field. Many of the concerns he's been speaking about have carried over to this new scientific frontier — particularly the ways in which ignorance tends to expand unrecognized along with knowledge. Practical and commercial applications of genetic technologies have surged ahead in recent years — think, for example, of the vast and increasing acreage planted in genetically modified crops in North America. At the same time, basic research in genetics is steadily complicating the once simple and straightforward story of how genes are supposed to work. A critical moment was the unveiling of the first draft, so-called, of the human genome in 2000.

BW: The fundamental genetic dogma ever since Watson and Crick discovered the double helix has been that one gene codes for one protein, and one protein creates one organismal trait — whether it's resistance or susceptibility to a particular disease or a propensity to go bald or whatever else it might be. Whether it was a relativity trivial trait or a very significant and serious trait, the fundamental dogma was, one gene, one protein, one trait. It was thought that if we could find out what gene does what — what gene creates what protein creates what trait — then we could do all kinds of things to actually improve our lot as human beings — our health or the milk production of cattle or whatever — by fiddling with the genes. On the basis of this central dogma, it was thought, it was expected that there'd be something like a hundred and fifty thousand or so different human genes in the genome. Instead of that, it was discovered that there are actually only something like twenty to twenty-five thousand human genes. The number has actually been coming down gradually, as time has gone on, so, instead of about twenty-five thousand, it's now down to something in the region of twenty thousand. This caused a big problem, because, if there were that few human genes, then you couldn't uphold the central dogma. There weren't enough different genes to create all the different organismal traits that we're supposed to be explaining by that determinist, linear sort of model.

DC: The revelation that people have less than a fifth the number of genes originally predicted is just one of a number of recent setbacks to the genes-explain-everything euphoria that followed the modelling of DNA. Subsequent genetic research has continuously added new layers of complexity and context to this simple model of gene action. Yet at the same time, commercial applications of genomics have rapidly expanded and biotechnology has boomed. This is the paradoxical reality that Wynne is now trying address. As in the past, what has interested him particularly is the way in which genomics is being presented to the public.

BW: The consistent question for me, throughout my career, has been the one I started with, which is, how is scientific knowledge being constructed and used in public domains? Now genomics, and bioscience and biotechnology generally, have come more and more into the public eye. Commercial interests are involved. Countries want to keep their economies competitive globally. Pretty well every country on the face of the planet these days wants to be a knowledge-competitive society, and bioscience is seen as the big golden goose that's going to lay lots of golden eggs. We just have to invest in it in the right way, and that means also investing in the social research that will get society to actually enthusiastically embrace all the innovations that are supposed to flow from it. That's a very common way of looking at things, not just among scientists but also among policy makers. Very often, what you find is that the social research agenda in this domain for people like me is defined in terms of delivering a quiescent public for whatever innovations science comes up with because that's crucial for the economy.

I can quote you reports from the European Union, for example, saying precisely that. I've had that said to me personally across a dinner table by government ministers in Britain. That's your agenda. That's what you social scientists should be delivering. Make sure that the public accepts what science delivers. Science is a public good, by definition, so whatever it comes up with . . . okay, there might be the odd mistake here and there, but the net result is benefit, social benefit, and you've got to make sure that that is appreciated by society. I want to say, no, sorry, I've got a few more questions. The scientific knowledge that is produced by science today is a function of all kinds of contingent social and historical factors. We've got to take responsibility for it, not just say that all this is coming out of some kind of nonsocial, completely natural fountain from which we enjoy an endless foamy flood of benefits. It isn't like that. In other words, the social research agenda is not only about the impact scientific knowledge has. It's also about how the knowledge is being produced.

DC: The question of how knowledge is produced takes us back to our starting point: Brian Wynne's recognition that science's purposes are always determined in a social context. Knowledge is potentially infinite. What we can attend to at a given moment is severely limited. There's always a question as to what will count as knowledge in a given context, and another about who will decide what counts. These questions are almost always properly political. That is, they require a judgment about what is good, a judgement which the scientist is no more competent to render than any other citizen. Wynne's examples at the moment are drawn from genomics. Contemporary research in genetics has replaced the image of the all-powerful gene with a picture in which genes interact with a complex cellular environment. Intervening in this complex environment inevitably involves decisions about what to attend to and what to ignore. This is explicit in the scientific literature he's been reading.

BW: When you look at some of the ways in which that science is described and the scientific strategies that try to make sense of that complexity, you can see that it requires you, by definition, to select.

You want try and model the thing you're interested in, and in order to model it, you've got to reduce the number of parameters you're trying to actually process and hold in play in your models. You've got to reduce the complexity. Some of the scientific papers that I've been reading say things like, we've got to select those parameters which give us potential therapeutic intervention pathways. Well, fine, you say, that's okay. That's a perfectly good social benefit. It's a worthwhile thing. But then you've got to notice that that means that we're systematically neglecting lots of other parameters and lots of other interactions that are going on in the complex biology of the cell. It might be that when we've developed the drugs out of this kind of therapeutic gaze on the cellular complexity, and then administered them to human beings out of a wish to help them, some nasty, unpredicted side effect may arise from the interactions that were ignored. Who's responsible for that? We didn't know about something that was going to create that side effect. Well, should we have known? Well, that's a question about how we posed our questions and about what caused us to focus only on certain parameters and not on others. There's a question of creating ignorance as well as creating knowledge. We create conditions in which we're looking for only one thing, so the question is always, for what purpose are we seeking this knowledge? We're pursuing knowledge in order to intervene. We're not doing it just because we want to know. We're doing it as technology. That's why my field calls this whole area techno-science, and not just science. It's not just innocent knowledge. And we've got to take responsibility for that as society.

How Reasonable Is Scientific Reason?

SAJAY SAMUEL

It is through our senses that we apprehend the world. The senses are also the way we are tethered to what is given, and we cannot find our nature unless we acknowledge our naturalness, which is shared with the world around us. Unless we remain tied to the world, our fingers, as it were, plunged into the soil, we will forever mistake our nature. Common sense is the gateway that links the world to my mind.

— Sajay Samuel

In 1543, Nicolaus Copernicus published *On the Revolutions of the Heavenly Spheres*, the book that displaced the earth from the centre of the cosmos. Ninety years later, in his *Dialogue Concerning the Two World Systems*, Galileo Galilei praised the achievement of his predecessor. Copernicus, he said, had made "reason conquer sense." Today, it is a commonplace that science requires us to renounce the evidence of our senses if we are to understand the true nature of things; the truth lies behind or beneath the appearances. But this loss of the senses has fateful consequences, according to Sajay Samuel, a professor in the Smeal College of Business at Pennsylvania State University. Without common sense, he says, science fills our entire horizon, leaving us no place to stand outside of science and no basis on which to judge what science produces.

Sajay Samuel's work on science and common sense began as a conversation that he carried on over many years with the late Ivan Illich. Illich was a critic of the power of professionals and experts. In books like *Deschooling Society* and *Medical Nemesis*, he challenged the idea that people living in a modern society must inevitably defer to those who claim authority on the basis of their scientific knowledge. Samuel encountered Illich nearly twenty years ago at Penn State University, where he had come from his native India to do graduate work in business administration. He was impressed by Illich's teaching. Friendship and then a close intellectual collaboration followed. After Illich died in 2002, Samuel continued the work they had begun together. He focused particularly on the question of how scientific judgments are to be judged. Is there, he wanted to know, a common-sense basis on which people can question the scientific knowledge that justifies the rule of experts?

Sajay Samuel and I have known each other, through our common friendship with Ivan Illich, for many years. For our interview, I set up my microphones in the library of his home in State College, Pennsylvania, and invited him to talk to me about science and common sense. He began by describing the general outlines of the inquiry he had been pursuing.

SAJAY SAMUEL

I began to wonder whether the distinction I was seeking between science or scientific knowledge and common sense would allow me to speak of the limits of scientific knowledge without falling into a kind of New Age romanticism, without falling into a kind of emotionalism, irrationalism. I wondered if a historical investigation into the distinction, if any, between scientific knowledge and common sense would allow for a reasonable critique of science — science which presents itself as coeval with reason and, therefore, allows no space for reason outside of science. If science is rational and science is knowledge, as it were, then the only critique of it is irrational. And if you refuse to critique it or you can't critique it, then you get objectified because scientific knowledge — we know this from our high school textbooks — is a kind of knowledge that's rooted in mathematics. But we intuit, all of us, that we can't be described in a formula or a theorem. One begins to sniff out this paradox: if I accept scientific reason as all of reason, then I become a theorem, which, when I bang my nose against the wall, I know is not the case.

DAVID CAYLEY

Science, in Samuel's view, creates a rupture between knowledge and experience. I know myself, when I bump into a wall, as a feeling subject, but science knows me as an object, a theorem, and science claims the crown of reason itself. The Romantic response has been to reject reason and reassert the primacy of experience. Samuel has taken a different tack. He wants to reclaim reason, and this has led him to investigate the ways in which people understood things before the invention of modern science. One of his first discoveries was that common sense formerly had a very different meaning than it does today.

SS: The meaning of common sense as it existed since Aristotle, for two thousand years, did not refer to the innate capacity of each person to know, but, rather, to a physiological dimension of human cognition. Common sense is the faculty that synthesizes the separate sensations that are picked up by your five senses and forges a coherent whole of that which you see, hear, smell, touch, taste. Very briefly, if I were to give a bit of a caricature, according to Aristotle and Aristotelianism, when I see a door, the door emanates its quality. The eye goes out and grasps that which the eye can grasp — each sense can only grasp that which is proper to it — so it apprehends, for example, the door's colour. But that's not the whole door, is it, the colour. There's more to the door: the size, the shape, the sound that it makes when you hit it, and so on and so forth. How do we get a coherent picture of the door? Different senses pick up different aspects of it, and these are synthesized in the common sense, which then allows us to make a judgment regarding the door, the key point being that *sensus communis*, or common sense understood in the Aristotelian way, is a faculty that permits judgment of things apprehended or the world apprehended through the senses.

Thus all ideas are a reflection of this original grasping of the world. It can't be cut, separated from the world. It can't come unhinged from the world. Your ideas, whatever ideas you have at the end of the day, via the synthesizing act of the common sense, are rooted in the world.

DC: There is nothing in the intellect, Aristotle wrote, that is not first in the senses. His saying became the common wisdom of the Middle Ages. What it means is that our ideas, however abstract, always draw on our sensory experience.

SS: One has to recognize the objectivity of the idea of ideas in the time before Descartes: ideas are not fantasies generated by the mind, for the mind, but are understood as inhering in the world. It is not the case, I think, for somebody like Aristotle, that justice is a figment of my mind. Justice is a property of the world. The idea of justice comes to me, for example, when I see two people sharing or cutting a pie. Seeing this gives me the rudiments of what is a proper or appropriate cut. Should you have more? Should I have less? Well, it depends. Who are you? How much do you need? And that's how we develop or come to the idea of justice. A classic case is when Achilles — this is before Aristotle, of course — when Achilles wants to distribute the booty obtained from a raid. He has everybody sit in a circle. The booty is thrown in the middle. Now we have to get into the idea of how to share this booty — a practical instance of justice in formation. In this Aristotelian tradition, common sense allows the intellect to be tethered to the world through the senses. It's the toll bridge, if you want, or it's the connection by means of which the world and the intellect are married.

DC: The marriage between the intellect and the senses comes to an end with the invention of modern science. Samuel follows a long tradition of historians of science and of scientists alike in making Galileo the archetype of the new attitude.

SS: Galileo starts with the proposition that the book of nature is written in number, weight, and measure; that is to say, when you look at a rose, it's not red. That's only what you see. That is a secondary quality.

The real rose is described geometrically. He falsifies your senses. The real properties of things are triangles and circles — mathematical, geometrical forms. It is a prior metaphysical commitment that allows him to make this move. He starts with the proposition that nature is number, and it is this prior metaphysical commitment that allows him to deny the reality of what he sees.

DC: Galileo falsifies the senses. What we seem to see is not what is there but what the mind constructs. Galileo's contemporary, René Descartes, takes the same view. As an example, Samuel describes Descartes' theory of vision.

SS: Descartes insists that there are atoms, little balls. The balls bump into each other. Their rate of spin — how fast they turn — generates a certain colour, and it is this matter in motion that allows you to see. Seeing is a result of little balls bumping into your eye, spinning fast, generating the colour — it's a caricature, of course — as opposed to the pre-modern account of seeing in which the object, the door, sends out its form, its qualities, which emanate, as it were, from the door. The object is not inert, matter is not dead. The door sends out its forms, and the eye goes out and grasps them, and that's the way seeing occurs. The door is not a figment of my imagination. The door actually is there, and it sends out its qualities, which I apprehend from the senses, as opposed to: there is no door; it's a bunch of molecules and atoms, little round balls bumping into each other, spinning very fast and generating my vision of a door.

DC: The key difference between these two views lies in whether our ideas are derived from things as they are or imposed by our minds. Are we of the same nature as our world — able to see the sun, as Goethe says, because our eyes are sun-like — or do we stand apart, constructing reality from data that have no inherent meaning?

SS: The mind gets cut from the world. In the older view, the world phenomenally presents itself in a certain form, in a certain way, you apprehend it, and the intellect forms conceptions of it; these are concepts of the world as it is. On the other hand, if you insist, as Galileo does, that what is real about the world is not what you see, what your senses grasp, but, rather, timeless geometrical forms of triangles and circles and squares, then the real is what you have imagined. You never see a perfect square in the world. That is something you cook up in your head and then impose on the world. The rational is treated as if it were real. You construct a square in your head, and then you say that the world is like a square or is a square, or that a rose can be described in terms of squares, circles, and triangles. This reversal seems to me of the first moment. You move from "concepts" which reflect the accord between the mind and the world to "constructs" which recast the world in the light of your ideas. And these ideas are freely created, unhinged from any thing, godlike.

DC: Mathematical ideas, Samuel says, are freely created by the mind. This does not mean, obviously, that mathematics has no purchase on the world. That mathematical ideas mysteriously correspond with the cosmos had been known since antiquity. What was new in the seventeenth century was the thought that these timeless forms could be applied to the living, ever-changing world of earthly nature. It took a bold step to get to this idea, Samuel thinks, because, during the Middle Ages, mathematical knowledge, particularly geometry, had been considered godlike.

SS: Geometry, geometric ideas are not derived from nature but, rather, emerge from the mind. They are bodiless and timeless. They are cooked up by the mind. The productive act of the mind, the imagination, in generating these forms is used as a way of understanding what God's creative power is like. God can make from within himself that which he thinks up, and the product of his thought is the Creation, so existence follows his thought. This leads me to a point that I'm hesitant about but I think is right: people like Descartes, Galileo, Bacon, and others insist on the geometrical nature of the world because geometrical forms are cooked up by the human mind and, if instantiated in the world, allow man to be creative in the same strong sense in which God is creative. It's a mimicking, if you want. I think if you ask the question, why does this metaphysical shift occur?, the answer has something to do with this immoderate aspiration to make the real in the light of the rational. The imagination creates mathematical forms. I insist that the world follows in the image of those mathematical forms and then recreate the world to suit. Since these mathematical forms lie buried under the phenomena, I must vex nature. I must pull out her secrets, which is this gesture of experimentation. I must stage the world in order to decipher the mathematical relations that I have invented and discover them in and through my experimentation. I can do that only by staging the world to make it hospitable to my interrogations.

DC: The first scientists imitated their conception of God as the supreme intellectual who thinks up creation. Just as God knows the world by having made it, they would know nature by remaking it. To get at this quality of artifice in the new knowledge that was to be obtained by experiment, Samuel uses the theatrical metaphor of staging. Almost all the terms he has been using — constructed, invented, imposed, cooked up — make a similar point: scientific knowledge is something we make, not something that is simply given to us. But isn't this scientific knowledge nonetheless true, one asks, since it manifestly works? If our mathematical key fits nature's lock so precisely, why keep harping on the artificial quality of scientific knowledge? Samuel begins his answer by drawing a distinction he finds in the work of historian of science Peter Dear.

SS: Peter Dear notes that, when we today speak about science, we speak about science in two ways. We refer to bodies of knowledge that tell us something about the world as it is. We also speak about science as an instrument with which to change the world, to improve the world — the vaccine, the bomb, the car. It has both this instrumental face, to use his language, and a natural, philosophical face — natural philosophy being the study of the way the world works and the way the world is. When you ask, why is science true? Why is a certain theory true?, the tendency is to say, because it works, because the plane flies, because the vaccine prevents disease, because the atom bomb explodes — those stand as proofs of the truth claims of science. If science were false, if the truth claims made by science were false, this vaccine wouldn't work. If you ask, why does this vaccine work?, it's because the science is true. There's a circularity in this justification: it's true because it works, and it works because it's true. Peter Dear calls it an ideology, and he calls it that in part, I think, because it can be falsified. Take the case of radio waves, which is one of his two examples. The prediction of radio waves in 1880, I think it was, by Hertz, based upon a scientific theory propounded by Maxwell regarding the ether. The atmosphere is composed of ether through which radio waves propagate. Well, the radio waves were real — the prediction was sound; it worked — but the theory was not: it was utterly false. The other example he gives is of navigators, even today, who use the old geocentric astronomy rather than the heliocentric astronomy. Again, you can have a perfectly false theory regarding what the world is, and it's useful. You can get things done. This unquestioned justification — why is something true? Because it works. Why does it work? Because it's true — can be easily falsified. And yet is held. We don't tend to question the connection between knowing something and making it. That which is made is identical to that which is true. This constructivism, knowledge through construction, can be understood to be the signature of modernity.

DC: Scientific knowledge, as Samuel has described it here, is creative, constructive, wilful, and, therefore, in a certain sense, godlike. These attributes make it quite unlike the common sense on which people formerly relied. What concerns Samuel is that science, historically speaking, doesn't just supplement common sense; it displaces it. Scientific knowing becomes the epitome of reason and the paradigm of all proper knowledge. The term "common sense" continues to denote sound judgment, but it also begins to evoke a certain ignorance of how things really are. Typical, in this respect, is Albert Einstein's often cited remark that "common sense is nothing more than a deposit of prejudice laid down by the mind in childhood."

Samuel wants to contest science's monopoly of reason. He would like to restore the dignity of common sense and restrict the application of science. Two distinctions are crucial to his case: mathematical knowledge must be distinguished from judgment, and experiment must be distinguished from experience.

SS: The distinction between experiment and experience is of the first moment. One has to be clear about this. From Aristotle to Hobbes, experience is understood as the consequence of repeated sensory impressions, which accumulate in memory. Only then can we say we have an experience. It is even today understood that "he's an experienced man" means that he has gone through something repeatedly. That's the sense in which we mean "experience."

"Experiment" can be a one-off event, a single event. Not only can the experiment be a single event, but it requires the staging of nature. You have to interrupt, cut, parcel, change, arrange so that you can study nature, so that you can interrogate it. This is not what obtains in the normal course. Experiment stages nature and then generalizes from that staging as if it were experience and anybody, anywhere, could repeat the procedure. That's not true. It takes a lot of apparatus to repeat many experiments. These experimental results, which are one-off events, single events or events repeated in highly stylized settings, by their very nature are extraordinary. But they are presented to us as evidence of the ordinary course of nature.

I make a distinction between experience and experiment: experience refers to what all of us usually will go through in interaction with nature in its normal course, while experiment is the consequence of the staging of nature, typically producing results that are not ordinarily visible but that are then rhetorically justified as being ordinary.

DC: What Samuel calls the rhetoric of science is the way science has been presented to the public from the time Robert Boyle first published his experiments in 1660 right down to the present. This rhetoric tries to persuade us that experiment should count as experience, that we should treat items of scientific knowledge as if they were as real as what we can taste and touch. Samuel would like to carefully distinguish between them, and he puts equal weight on a parallel distinction between mathematical demonstration and common-sense judgment.

SS: Numbers give answers in the way a computer does: zero, one; true, false. If I say this table is six feet long, then either it is or it isn't. There's only one answer. Even when people talk about statistical variability and so on, there is still one correct answer. We can't fathom it to the degree of precision that we want, but, nevertheless, there is one correct answer. All numerical measurements are of that character.

There is in Plato, for example, another kind of measurement. It is the measure — and it is a measurement — of too much, too little or just right, but that second class of measurement is an act of unquantifiable judgment. It depends. It depends on the situation. It depends on the human intention. It depends on who's asking the question, for what purpose. The answer is not an unambiguous six and a half inches.

This distinction between mathematical measurement and what we call a judgment regarding measure — is there too much or too little? — comes to the fore in what we're going through or what we're facing now, for example, in questions of global warming. Global warming is a scientific hypothesis. It's either true or false. Either the carbon is .2 or it's .5, but whether that is too much or too little is a matter of judgment. It depends. It depends on the human intention. It depends on human purposes, et cetera. There is no scientific answer to that question.

More deeply, we can ask, should we be understanding nature in this way? Now I'm not speaking just of global warming. Is there too much or too little science or scientific ways of knowing? It's a perfectly legitimate question, and there is no scientific answer to that. This means, in principle, that experience and judgment are superior to scientific knowledge in rank, because we can always ask of science, is there too much of it going around?, and there is no scientific answer to that question. The muteness, the silence of science when the question of its own legitimacy is posed shows that there is a higher authority in this regard: common sense. I don't mean mystical authorities. Simply common sense. It is this point that I want to underline. Our ideological slavery to the identity between science and reason — scientific reasoning is the only way of reasoning — can be upended by noting this one fundamental flaw: science cannot answer the question of its own legitimacy. This provides a lever by which to free ourselves from our ideological, blind identification of science and reason. Asking about the legitimacy of science is an extremely reasonable way to recognize the muteness of science and thus put it in its place, as it were.

DC: Samuel wants to establish two crucial distinctions: between experiment and experience, and between mathematical demonstration and common-sense judgment. By means of these distinctions, he thinks that science can be, as he says, put in its place. It can be made to stop masquerading as reason itself and made to answer to an authority outside of science. Otherwise, we will turn endlessly in the same circle in which the answer to the problems generated by science is always more science. Why does putting science in its place matter? Well, first of all, because science, in a broad sense, now completely dominates our public and political life, defining both what we think about and what we do.

SS: What is the subject of political discourse? It's global warming. It's Kyoto. It's cars. It's economic distribution and unemployment. It's sex policy. It's foreign aid. It's flows of trade, et cetera, et cetera. First, this subject matter of political discourse is constituted by objects made from economic science, biochemistry, et cetera, et cetera. Second, in thinking through these matters, these subjects, the apparatus or forms of thinking that are brought to bear are themselves scientific. Political questions are dealt with through public policies, and public policy is an arm of political science. What passes for politics today — both in its subject matter and in its resolution, if you want — is framed in scientific terms. The polemical way to say this would be that public policy consists of experiments on the *polis*, scientific management of people.

DC: Modern political life is thus scientific in two senses. First, its contents are framed scientifically. Even an ordinary newscast will refer familiarly to things that might as well be angels or gryphons, as far as most of the audience is concerned: a critical shortage of radioisotopes, let's say, or an adjustment of so many basis points in the interest rate. Second, people are managed scientifically by the arts of polling, public relations, health promotion, and so on. That's what state policy now is, in large part, Samuel says.

SS: If you investigate the term "policy," it comes out of "police" and the sciences of police, which exist to order the political space. It becomes the armature, it becomes the vector through which you manage the political. That's the role and purpose of public policy. Now, it seems to me, insofar as it is informed by science, social sciences, it is nothing but the scientific experimentation on people.

DC: If one takes your definition of what Galileo is up to — recasting the real as the rational — it seems very hard to object to what you're now saying about how modern politics operate, which is to bring people in line with rational designs for their organization, for their improvement, and so on. How else would we define it?

SS: Aristotle has an interesting understanding. He says, "Man, by nature, is a political animal," and he argues that all animals, man included, have voice. Voice gives expression to pleasure and pain. A cat's meow, when you slap it on the rump for sleeping on your bed, is aggrieved in a way that its meow is not when you feed it. That we share with animals, but only man has speech, and speech is the way in which man expresses his concern for justice and injustice, what is good and what is bad, what is fitting and what is not fitting, what is proper and what is improper.

The scientific management of man — if we agree on what science is — is inhospitable to questions of just and unjust, proper and improper. Those are questions of judgment, requiring you to understand what is too much and too little, what is in the mean. In fact, the scientist himself admits this. He says, that's the domain of values. But values are utterly irrational, and that's why I oppose this whole way of setting up the question. The distinction between science and values is false inasmuch as it leaves no room for a reasonable critique of science. Values become a subjective, irrational commitment to decision. Why do I do something? Because I want to, but I have no reason for it. Coming back to my point, the kind of speech that characterizes the political animal for Aristotle finds no soil in the world of public policy.

DC: Samuel's response to a politics that uses scientific techniques to manage scientifically framed issues is to turn his back, not out of indifference, nor out of intellectual pride, but because he believes that only by rejecting a politics immune to common sense can true political judgment be reborn.

SS: There is a first step, it seems to me, in recognizing the inhospitability of what passes for politics today to what is properly considered political speech — namely, questions of justice and injustice, what is appropriate and what is not. That first step, recognizing the inhospitability, is to establish some distance and to refuse to engage with questions that are posed in purely scientific terms. We have to ask whether or not there is too much science and scientific ways of talking and speaking in politics — that's our first step — rather than what we're invited to do now, which is simply to take sides on so-called political questions. Either you're for GMO or against GMO; either you're for global warming or you're against global warming; either you're for outsourcing or you're against outsourcing. Your credentials as a so-called political animal are given by this blind decision as to which side of the debate you're going to come down on. But this is done in perfect incomprehension. We can't understand or comprehend these questions in a commonsensical way. We take on faith, at the end, the truths regarding these questions. Rather than engaging in what, traditionally speaking, is no more than apolitical noisemaking — beating our drums, standing in lines, fighting for or against this or that — I would suggest that we first ask whether we can we step back and get rid of a type of speech which does not lend itself to what is primarily political — namely, questions of justice and injustice — and that does not allow us to make a judgment as to whether something is too much or too little?

To make it concrete, I can't get into the discussion about climate change — and I've read quite a bit, though nowhere close to what is perhaps necessary. I don't want to deny for one minute that man has despoiled his home, where we live. We shit in our own backyard. There is no question of this. However, the conundrum we face is the following: as this becomes a scientific reality, the solution is going to be scientific and will entail the scientific management of people, whether it is through pollution credits, whether it's through increased gas prices, whether it is through redesigned urban spaces — whatever. The idea that man has gone too far with industrialization, which is the proximate cause today for all of these troubles, was understood long before and without any science. The solution then was also non-scientific and commonsensical: saying, this is too much. There is no need for this. We are going too far. The routine disregard for common sense and privileging of science is precisely what has led us here. I think there is no way out except through a return to common sense. Otherwise you end up with more scientific management. You don't end up anywhere else than where we are already.

DC: Samuel's reticence vis-à-vis climate change as a political issue has two foundations. The first is reluctance to assume the distanced point of view of the climate modeller, who surveys the earth, as it were, from a great height, abstracting from every weather and encompassing every age of the earth. The second is his feeling that the discussion of climate change is not usually about the proper limits to human activity, about how far we ought to go in our exploitation of nature, but almost always about how far the biosphere can be safely pushed before it pushes back — in other words, about what we can get away with.

SS: Let us ask, what is the purpose of engaging in these climate change models? Mind you, I have no doubt — and I want to underscore this fact — that the earth has been despoiled through industrialization. All I have to do is to walk in my home in India, for example, and take a breath of what passes for air. It's manifest. It's common sense. I'm not questioning that. Let's go back to the issue: why these climate change models? It seems to me that the question that is being posed is, how far can we despoil the earth before it takes its revenge? It's an eminently practical question. How much can we continue to destroy before the earth bites back? But it seems to me that the prior question to ask is, are there natural thresholds that even those of the meanest capacities can see should not be crossed? The moment you exceed what you are capable of, the moment you move from a bicycle to a car, you've broken a certain natural threshold. In breaking that threshold or limit, not only is it likely that you will despoil the earth, but, more importantly, it is likely that you will lose the centre from which you can see that you're despoiling the earth. I think that is terribly important. The more credence we give to the necessity of posing questions scientifically and eliciting scientific answers, the more we buy into the necessity of being scientifically managed.

DC: What politics should be about is what every citizen can understand and experience. Otherwise, Samuel says, only the expert is entitled to an opinion, and people are deprived of any opportunity to exercise judgment. How many citizens are competent to judge the appropriate atmospheric concentration of carbon dioxide? When technical questions dominate public discussion, the majority is disenfranchised. In consequence, ignorance and incompetence intensify along with expertise.

SS: In the Age of Reason, Enlightenment, techno-scientific rationalism, this apogee no civilization has reached in the history of man, there are many varieties of ignorance. Marc Augé, the French anthropologist, notes that Americans are finding other Americans and Europeans are finding other Europeans as strange as once the British found the Indians and the French found the North Africans. Americans have begun to do an anthropology on themselves, so what was once reserved for the strange savage is now being applied at home. There seems to be a certain obscurity among Americans regarding themselves. Not to speak of another variety of ignorance that is expressed in identity politics and the search for identity: I don't know myself, I'm going to figure out who I am — a certain obscurity regarding oneself, even. More than that, we're surrounded by techno-scientific objects which, strictly speaking, have constructed for us a world of magic, where things work but the cause is occult or hidden. I go to the ATM. I put in a card. I punch in a certain secret code, and it spits out money. How and why, I have no idea. I was in France, and it happened there, too. I call my mother, and even though, somebody tells me, there are no wires, there is nothing, it just bounces off the air somewhere, yet she hears my voice. It seems that we're living in an age of magic in the strict sense, surrounded by objects that work but whose causes are occult. There's a certain obscurity in how things work.

Finally — and I find this particularly telling — the production of science has gotten so complicated that one scientist, or one class of scientists, doesn't know what the other guys are doing. This came to a head, for me, when David Ruelle reported in *Chance and Chaos* that the solution, or the purported proof, to the so-called four-colour theorem occupied five thousand pages, and no one mathematician could comprehend it — an obscurity even at the heart of mathematics. That is to say, rationalism has reached a stage where ignorance seems to be rife. If that is not a call to rediscover or to find out a reasonable critique of techno-scientific civilization, I don't know what is. In this civilization, which enjoys what are considered to be the highest and most prolific fruits of reason — the byword of the Age of Reason was "dare to know" — we have reached, three hundred years later, an almost systemic ignorance. Nobody knows very much. Everything seems to be working. We are more and more made to behave irrationally. When I ask, how does the ATM work? Why does this work? Why does that happen?, answers are very hard to come by, and yet I'm expected to act. I'm expected to sign up, do things. We're in the midst of a world where action is based on unreason.

DC: Science, Samuel keeps stressing, cannot judge itself, cannot find its own proper limits, cannot tell us what is enough. For him, it is folly to answer each failure of science by calling only for a more subtle or a more comprehensive version of the same thing. What we urgently need is a language in which all citizens can discuss the questions of what is good for us. He proposes to revive an account of reason that can bring science under common judgment. But how, I asked him, can this even be imagined? Our public and private lives teem with scientific objects, and our speech is full of half-understood migrants from the sciences. How is a return to common sense conceivable? His answer appealed to what Canadian philosopher George Grant once called "intimations of deprival," the feeling that something is missing in the modern, something we might rediscover were we to bracket our constant recourse to science.

SS: One way to get distance from scientific terms and certain taken-for-granted terms, certain words, is to begin to undertake the exercise of finding out what they displaced. In attempting that work, I rediscover a residue, a remnant, a "rest," something left over, that allows me to express, however incoherently, however gropingly, a dissatisfaction with the scientific terms, a recognition that they don't quite address what we are asking. There is a gap between what the term purports to say and our acknowledgment that it doesn't quite say it, and that gap provides the fillip, the inducement, to undertake the necessary historical investigation. You ask, what is lost through our being saturated in these scientific terms? I think the best way I can say this is through a quotation. The book I'm quoting from is *Hermeneutics as Politics*, by Stanley Rosen. He writes, "If knowledge is enlightenment and science is knowledge, it follows that to be enlightened is either to endure self-ignorance or to undergo reification." If knowledge is enlightenment and science is knowledge — this is that equation we spoke of earlier — then to be enlightened is to suffer self-ignorance, because science cannot reach me, or to become an object, undergo reification. Why does it matter, you ask, that we are saturated in scientific terms? If it is through scientific terms that we understand ourselves, then we treat ourselves as objects. Or, if we recognize that scientific terms cannot help us understand ourselves, then we are forever obscure to ourselves, we are ignorant. Insofar as that scientific milieu does not change, we are condemned to ignorance. Man, the rational animal, finally has become an animal. This is why the stakes are, it seems to me, so high.

DC: At the beginning of our conversation, Samuel pointed out that science denatures our senses. It shows us a world that is not what we sense it to be. Consequently, it is in a return to our senses that he sees the way out of the dilemma he has just described: under the rule of science, we must either come to see ourselves as objects or remain forever obscure to ourselves. It is through our senses that we belong to the world and the world to us.

SS: If it is through our senses that we apprehend the world, the senses are also the way we are tethered to what is given, and we cannot find our nature unless we acknowledge our naturalness, which is shared with the world around us. Unless we remain tied to the world, our fingers, as it were, plunged into the soil, we will forever mistake our nature. Why does common sense matter? Common sense is the gateway that links the world to my mind from the outside in. If I speak of natural thresholds as limits, this tethering is central or key. I have to remain tethered to the soil to know, to recognize a natural threshold. It is precisely because we've been uprooted and freely floating, for the most part, that we don't grasp anymore, in a carnal, sensible way, the difference between walking and cycling and being FedExed. When you get into a car, you are made into a FedEx package, just as I am. The fact that we think of walking and driving as no more than alternative modes of transportation is an index, is a clue to how far we have come untethered, how uncommonsensical we have become.

DC: And you think that a first step, which anyone can take, is to recognize the difference, even if one can't initially do anything about it?

SS: I think the demand to act, posed with an almost blackmailer's urgency — either you act or you die — is a seduction into further losing oneself. There is ample time. Even if I can't do anything, to think right, to attempt to think right, to study, to work, to clarify is a form of doing. To rediscover what it has taken three hundred years to uproot can't be done in a day.

Everything Speaks

DAVID ABRAM

Whether we are scientists or slackers, whether we're farmers or physicists, we've all been born into a culture that profoundly conditions and constrains our senses. I think that this culture deeply conditions all of its members to not really notice the enigmatic wonder of the world that surrounds us on every hand, to not really notice that that world is actually there.

— David Abram

From time to time, researchers test the public's understanding of science. The public, predictably, turns out to be woefully ignorant. Twenty percent think the moon is made of green cheese, thirty percent think an electron is bigger than a molecule, and so forth. For David Abram, this demonstrably shaky grasp of the details misses the point. He's the author of a widely read and much praised book called *The Spell of the Sensuous*, and he thinks we are conditioned by scientific understandings at a much deeper level. The main effect of this conditioning is to make us distrust our senses. For citizens of the republic of techno-science, he says, the real world is not the one we can touch and taste. It's the one that is disclosed by particle physics or radio astronomy.

In a recent essay, Abram recalls how this conditioning occurred in his own life by looking back on the high-school science education he experienced as a nature-loving boy growing up on Long Island in the 1970s. He begins with the earnest physics teacher who announced to the class that the apparent solidity of the table on which he was sitting was an illusion, the table actually being composed of empty space shot through with tiny whirling particles. Then there was biology, taught entirely in a laboratory-like classroom with no reference to the living world surrounding the school. There, the teacher explained that animal behaviour is "programmed" in each creature's genes, that the cardinals, thrushes, and blackbirds, whose songs had entranced young Abram, were "not really saying anything." They were just feathered automatons, obeying coded instructions.

Abram took this familiar style of scientific education unusually hard. The singing, speaking world disclosed by his senses had mattered to him. It was a blow to find his teachers treating this world as nothing more than a hollow and mechanical appearance, a will-o'-the-wisp emanating from the more fundamental realm disclosed by mathematical formulas and high-powered instruments, but what he learned in school was, he thinks, typical. In his view, most of us come to assume, one way or the other, that the world as science knows it is somehow more real than the world as we experience it. For him, this gets things entirely upside down: the real world is the one we can touch and taste, hear, smell, and see. Science is only its abstract or diagram, unintelligible except by reference back to sensory experience.

David Abram wants to snap us out of our techno-scientific trance by restoring the primacy of the senses and putting science in its proper place as a powerful and useful but still secondary form of knowledge. Speaking from a radio studio in Santa Fe, New Mexico, near his home, he told me a little about how he came to see that modern techno-scientific civilization had lost its grounding in the soil of the sensible and to understand that it is by our senses that we are in the world and the world is in us. One of the ways he learned was through the profession he first adopted to put himself through college: sleight-of-hand magician.

DAVID ABRAM

As a result of practising this craft, magic and sleight-of-hand magic, I became very, very interested in perception and in our ordinary sensorial experience of the world. Perception, it turns out, is the medium for a magician, as pigments are the medium for a painter. Magicians work with this very malleable and quite mysterious element that we call perception, perceptual experience. Whether one is a contemporary sleight-of-hand conjurer or a traditional, indigenous shaman or medicine person, one is still working with the fluid quality of perception itself. Magicians are those who are adept at altering or shifting the accepted, conventional perceptual experience of their community, whether they do so in order to enter into rapport or communication with another shape of intelligence, such as a spider or a wolf or a whale, or whether, in the contemporary world, magicians practise somewhat the same thing, often just for entertainment purposes, which is a kind of falling away from the deep meanings that have always been there in the practice of magic. It was from my craft as a sleight-of-hand magician that I became so fascinated with sensory experience and with how our eyes, our ears, our skin feel and make contact with the larger world around us, how they bring us into a kind of felt reciprocity with the living world that we inhabit.

DAVID CAYLEY

One of the things that brought the world alive in this way was a year Abram spent in Asia, comparing notes, so to say, with traditional magicians. He had practised his sleights in clubs and restaurants in the US, worked his way around Europe as a street magician, and spent several months as part of maverick psychiatrist R.D. Laing's Philadelphia Association, where he explored the use of magic in psychotherapy. Then he decided to travel to Indonesia and Nepal, where his abilities gave him an "in" with the shamans, or medicine people, of those cultures. The experience so changed him that it was a shock to come home.

DA: When I returned from my first travels among traditional, indigenous peoples and communities, when I returned to North America, it was quite a warp in my own felt experience of things; suddenly I could no longer feel the activity of the shapes around me, of the ground itself. In the course of living with traditional magicians, I found my senses coming awake in a way that they never had in my own culture since I had been a child. I had grown up into a culture that defined the sensorial surroundings as basically a set of inert or inanimate or at least determinate objects and objective processes. My senses had become blunted. My eyes had glazed over, and I had been living more and more in a set of abstractions.

In the course of living with and conversing with traditional peoples and, in particular, these traditional healers and medicine people, I had to learn this way of speaking that allowed each thing its own active agency, its own life. As I took on this way of speaking, my own senses became much more awake to where I was, and I started noticing the outrageousness of the real world that surrounded me on every hand. Stones, small plants, gusts of wind, patches of lichen slowly spreading on a boulder — everything became fascinating, strangely weird, each in its own way, because I was now listening and looking at things with a kind of expectancy, listening for what they were up to, no longer just defining them as inert or determinate processes. I was sensing that there was an otherness present in every elemental aspect of the earthly world, and it seemed to me that this expectation of otherness — that everything has its own interior spontaneity, that each thing has its own life — this way of speaking is simply a way of talking in accordance with one's senses and with one's direct sensory experience of the world. It's a way of speaking that holds one's senses awake and aware of the immediate surroundings.

DC: The world is alive, Abram says, but our senses grow dull as a result of our definitions. The world cannot surprise us because we live in what he calls a "set of abstractions." The senses are overridden by more general schemes and categories. This tendency is built into contemporary culture in many ways.

DA: Whether we are scientists or slackers, whether we're farmers or physicists, we've all been born into a culture that profoundly conditions and constrains our senses into very specific modes of encounter with the sensuous terrain. I think that this culture deeply conditions all of its members to not really notice the enigmatic wonder of the world that surrounds us on every hand, to not really notice that that world is actually *there*. We've defined away all of the meaning in the sounds that surround us. The voices of the birds and the other sounding animals are not really voices. Certainly, they're not saying anything; they're only making automatic sounds programmed in their genes. It's as though other organisms are just automatons, and some computer programmer came along and inserted new software into them, and that's why they are acting as they do. Hence there's no mystery there to compel our ears to listen in deeper. There's nothing being said by those birds, and so our ears have become somewhat deaf to anything that does not speak in words. Our eyes have become rather blind to anything that is not human or of human invention because we have grown into a collective discourse that casually speaks of the rest of animate nature as a set of objects and objective processes. There's no creativity there. The movement of a crow or of a spider is not a creative response to what surrounds it in the immediacy of the moment, and so our eyes, yes, glaze over, and we become more or less blind. I think this is our common inheritance today. Whether we have been steeped in the sciences or whether we've paid them no attention whatsoever, it's part of the collective discourse of our time, a way of speaking that deeply influences our ways of seeing and of hearing, even of tasting the world around us.

DC: Abram argues that the senses of modern persons have been conditioned into a kind of complacency. Other beings don't speak to us because we have decided in advance that they have nothing surprising to say. This expectation arises from what he names a "collective discourse and a common inheritance." It holds all of us, whether or not we are versed in the sciences. Yet science, in a very broad sense, is its source because it is science that has been telling us for centuries that the world is not what it seems to our senses.

DA: I think a primary element of the scientific background discourse of our time is the teaching that we should not trust our senses. Every child is born with a kind of exuberant sensorium that wants to participate in every aspect of the sensorial surroundings. But fairly quickly, in elementary school and as we proceed through our educational system, we learn and take on in all sorts of ways, both explicit and implicit, the idea that we should not trust our senses, that the senses are deceptive, that the senses lie. It's a taken-for-granted aspect of the modern moment.

DC: The roots of this distrust of the senses, in Abram's view, lie deep in the history of science. In fact, he thinks, it virtually defines what we call the "scientific revolution."

DA: The key aspect of the scientific revolution was this new distance and detachment from our felt sensorial experience. The Copernican revolution, with its outrageous disclosure that the earth is moving and the sun is not moving around the earth, flew in the face of our direct sensory experience of the sun as this being, arcing overhead through the sky, slipping down beyond the horizon in the evening, and then climbing out of the eastern ground every morning and arcing through the sky again. This was our sensory experience of the sun for eons, and we suddenly took on this new way of speaking and of thinking. It said that this is not true, and, hence, you should not trust your senses. The truth is hidden behind the scenes. It is what the experts, with their access to very high-powered instrumentation, like the new telescopes of the time, must mediate to us. The truth of the world is written, as Galileo said, in the language of mathematics, and our senses bring us a deceitful story of how the world moves. The scientific revolution brought this tremendous distance from our senses. It's not well enough recognized that René Descartes' severance of the thinking mind from the body, his cutting of *res cogitans*, or thinking stuff, from the mechanical body, his detachment of the thinking intellect from the sensing body, was done in order to set up the ontology that was required by this new state of affairs: we should not trust our senses at all, but should really give ourselves over to the intellectual study of the world as it is hidden behind the scenes.

DC: Descartes' cut between mind and matter was, in a sense, a defensive operation, Abram says. It protected the mind from the deceits of the body and of nature. At the same time, this attempt to locate humanity outside and above nature also had deep religious roots. Popular accounts of the scientific revolution have often stressed the tension between the church and the new natural philosophy of Copernicus and Galileo, but Abram believes that, in a deeper sense, the new science was a renewal of long-established Christian habits of thought.

DA: So many of our comrades and colleagues speak of science as having displaced religion, but that seems very mistaken to me. So much of the world of modernity, and particularly the scientific discourse of our time, is a continuation of certain deeply religious assumptions, many of them Christian, regarding the distance of the human from the rest of nature, as well as the fallen character of the natural world. Although we no longer, of course, speak of material nature as "fallen" and "sinful" and "demonic," we now speak of it as "inert," "mechanical," "determinate," and, in many ways, "dead." But it seems to me that this is just a continuation of the same deeply religious prejudices, only translated into a new, much more modern idiom.

DC: Modern science thus continued the essentially religious theme of humanity's alienation from nature and nature's alienation from God. The new natural philosophy of the seventeenth century, in Abram's opinion, also sought an explicit accommodation with official religion. During the Renaissance, the experimental practices that were later characteristic of science were still intertwined with alchemy and magic. Giordano Bruno, burned at the stake by the Roman Inquisition in 1600, is a typical figure, a protoscientist and follower of Copernicus, on the one hand, but still, as we would say today, an animist, accused by the Inquisition of "dealing in magics and divination." The church was threatened by people like Bruno, with their belief in an active and spontaneous nature, but easily made its peace with the image of nature as a lifeless clockwork.

DA: The modern practice of science was in search of a new lingo that would disarm the wariness and the antagonism of the church. Mechanism — this whole metaphorics of the world as a vast machine —

was the perfect discourse to come in at this time, and it was grabbed onto by many of the early practitioners of science as a way of speaking that enabled their researches while, at the same time, implying that, well, if the world was a big machine, somebody must have made the machine. There was still this need for a divine source radically transcendent to the material world, a divine inventor, if you will. As mechanism became its discourse, modern science found itself in a much more fluid and easy relation with the church of the time and no longer in a kind of very uneasy and antagonistic relationship. I think that's a key reason why the discourse of mechanism swept through modernity and has become a taken-for-granted lingo — so much so that most people don't even recognize that it's a metaphor, because, of course, the world is not a machine. It wasn't built from outside. It seems to have been born of itself out of a cosmos that is in a continual process of creation, but of self-creation. I think that's worth pointing out.

DC: The mechanical philosophy fit in very comfortably with an authoritarian religion in which God was conceived as a mighty ruler and lawgiver. The church held that ultimate reality was available only through its mediation, and science made a similar claim about nature. The Bible tells you how to go to heaven, as Galileo is supposed to have said; science tells you how the heavens go.

DA: There is a sort of tacit alliance between the mechanical description of material nature and a very conventional Christian understanding of nature as a secondary realm, fallen away from the real source that is radically transcendent to the sensuous world. Even those brothers and sisters within the sciences who speak of themselves as avowed atheists nonetheless tend to view the material or sensuous world around them, the world of our direct experience, as a kind of secondary dimension, derivative from a more primary realm, and to think that this world that we experience is not nearly as real as that hidden dimension of subatomic particles, of quarks and gluons and mesons. Other researchers will say, yes, but of course our experience of things is also caused by happenings unfolding deep in the nucleus, not just of the atom, but deep in the nucleus of ourselves, in the genome and in the interactions between various strands of DNA, to which we gain access again only with very high-powered and expensive equipment and instrumentation. In these and many other ways, modern scientific discourse keeps telling us not to trust our senses and our direct sensory experience of the real, but to assume that the world we inhabit and experience is really explained by dimensions hidden elsewhere. This is, in so many ways, a perpetuation of the theological presumption that material nature is itself to be explained by reference to a divine source hidden entirely beyond all bodily ken.

DC: Science, at the time of which David Abram has been speaking, was a new philosophy, struggling to establish its legitimacy. By the turn of the twentieth century, its assumptions had become the new common sense, its view of nature so much taken for granted that people no longer even noticed that mechanism was only a metaphor. Science came to dominate modern societies, and it was this domination that gave rise to the philosophical school that has most inspired David Abram: phenomenology. Its founder was a German philosopher, Edmund Husserl, who lived between 1859 and 1938.

DA: Husserl, as the founder of phenomenology, was recognizing that the sciences had, by the early twentieth century, become so thoroughly estranged from the world of our direct experience that they were threatening, by their ongoing researches, to destroy that world; he was also recognizing that they had been taking us humans further and further from our felt experience of things. He set about developing a philosophy that would serve as the ground for the other sciences by being a science of perception, a science of experience itself, from which the other more abstract sciences could launch themselves without then necessarily setting themselves in opposition to our felt experience; phenomenology proceeded as the investigation of the world as we experience it. Prior to reflection, prior to the thematizations of mathematics and of science, how do we encounter the world in its immediacy when we just pay attention?

DC: Edmund Husserl wanted to recover an experience of things that had not already been cut and dried by the analytical mind, but, in Abram's view, he still held humanity aloof from nature in certain crucial ways. With Husserl, the consciousness or awareness to which the world shows itself remains outside nature, independent of the body, even though the body is the location of this awareness. He was still, in this sense, a Cartesian. It remained for a student of his, the French phenomenologist Maurice Merleau-Ponty, to take the next step and argue that the body is not just the location of consciousness; it is consciousness itself.

DA: Merleau-Ponty made a radical move to step out of Cartesianism entirely and to say that, well, the very one that is experiencing this consciousness that Husserl was alluding to, this bare awareness, is none other than the body itself. The body is the being of awareness, the body is aware, the body is the subject of experience, the experiencing being. This was a very radical move to make. In some ways, Merleau-Ponty had to develop or invent a whole new way of speaking so he could begin to step out of the Cartesian predilections of our contemporary discourse, so he could open up another way of speaking that would not tear us continually out of our senses and carry us out of our bodies.

DC: With Merleau-Ponty, we arrive at the project that David Abram has taken up and carried forward in his own work. Merleau-Ponty overcame the separation of mind and body in Cartesian philosophy and relocated mind in nature. Our thoughts are the world's thoughts, he said, our flesh, the world's flesh. If I think the sky is blue, then — to take an example from Abram's writings — the sky is equally thinking itself blue through me. If I speak *of* the world, then the world also speaks *in* me. In Merleau-Ponty's view, even language, so often taken as the crown and seal of human uniqueness, is not our exclusive possession but belongs to the world to which we are giving voice. This is an aspect of Merleau-Ponty's thought which Abram has been particularly interested in extending and enriching.

DA: One of the discoveries that was central to Merleau-Ponty's phenomenology was a recognition that, to the sensing and sentient human animal, it's not just that everything is, in some sense, animate and alive, but also that everything speaks, that any sound can be a voice, that any movement can be a gesture, a meaningful expression, that expressiveness is a property of the world itself. Everything around me has at least the capacity for meaningful speech. Even the hum of the fluorescent lights in the ceiling overhead is a kind of voice because there's a kind of meaningfulness present in sound itself. It's not an articulate meaning, necessarily. But take the particular sound the tires of a car make, as they whoosh along wet streets in a rain at night. That's a sound that affects the listening organism of our body. It slips us into a certain mood. Just as the songs of birds affect our mood or affect our state of mind, sounds influence us, and we respond. In some sense, our own human languages were all born in a call and response with the world of sounding, expressive shapes that we inhabited. It's quite clear, for instance, that our indigenous ancestors, our hunting-and-gathering forebears, were quite dependent on their ability to listen in on and learn from and even mimic the sounds of other animals. In order to draw those animals close, one sometimes had to make utterances that were very much like the sound of those partridges or the cries of those hoofed beasts that you wanted to get close enough to kill. Of course, this is before we had anything like guns. You had to get quite close to another animal if you wanted to bag it for dinner. For that huge stretch of our existence as *Homo sapiens*, we were in a very close rapport and a very intimate and attentive attunement with at least those other animals on which we depended for our sustenance. Our own languages would seem to have been deeply informed by the calls and the cries of these other animals, as well as by birdsongs and even the sound of the wind in the willows, because this too would seem to be a kind of a voice; each entity has its own eloquence, and our speech is just our part of a much wider conversation.

DC: What does that say to the view of scientific linguistics, that our languages are largely arbitrary codes?

DA: That's how you have to think of language if you're going to continue to conceive of language as an exclusively human property. You have to separate out all of the felt meaning in our sounds and just focus on language as a code wherein each term rather arbitrarily stands in as a sign for a particular meaning. But, of course, language is not strictly a code. All of our terms carry a sort of a poetic resonance within them that we can never fully cut away. I was thinking recently, as a simple example, of our words "thunder" and "lightning" — our words for the flashing slash of jagged light that cuts through the sky before we hear the thuddering, shuddering sound we call "thunder." Suppose one tried to reverse those two words and spoke of the sound as "lightning." The relation between sound and meaning is supposedly arbitrary, so let's call that jagged slash of light "thunder" and that deep, rumbling sound "lightning." Well, it wouldn't hold. You could try and practise it for a few days or a few weeks within your community, but pretty quickly it would reverse itself again because there's something about the word "thunder" that echoes the sound, not in any immediately obvious, purely mimicking manner, but it feels right. Certainly, it feels a lot more right than speaking of the sound as "lightning." It's not by chance that the words that we use to speak of the sounding speech of a river as it washes through its banks are words like "wash" or "splash" or "gush": the sound those words all share, "sh," is the very sound that the water speaks as it rolls over the rocks. Our human languages have been informed by many other sounds and voices besides those of our own kind. It follows from this that, as we dam up all of our rivers and clear-cut more and more of our remaining forests, as there are fewer and fewer songbirds in the surroundings due to the destruction of their wetlands and their wintering grounds in the tropics, our own human languages lose more and more of their meaningfulness, because they're no longer informed by and influenced by the splashing speech of those dammed-up rivers or by the cadences of the warbler and the wren.

DC: Language will diminish if we silence the world around us or replace it with manufactured sounds. More and more of his contemporaries live, Abram thinks, in a kind of technological enclosure or techno-scientific second nature. Scientific abstractions replace experience and digital technologies project us into what he calls "bodiless spaces." His response is to keep reminding people that, underlying both science and technology, there is still a "more primordial reciprocity" between the human organism and its world.

DA: Today we are surrounded by a kind of cocoon of technologies and often seem to have very little direct access to the more-than-human field of life. Everything seems to be an artifact. Everything I encounter in this room with my gaze seems to be an ingenious invention of my own species — until I look closer, until I realize I'm still breathing air as I'm sitting here, and the air that is moving between me and this microphone still contains oxygen which has been breathed out by all the green and rooted folks that surround this studio, by the grasses and the trees out there. I'm breathing in what they're breathing out, and I'm breathing out what they're breathing in. I'm still under the influence of gravity as I sit here, and the gravity that holds my body to the body of the earth is still as great and strange a mystery as it ever was. It was an uncanny mystery to our brothers and sisters at the dawn of the scientific revolution, but once it began to be spoken of as a law, the law of gravity, as soon as we speak of it as a law, then we stop noticing it. It's just happening automatically, so there's nothing very mysterious there. But what is this mystery? We define it even today as the mutual attraction between bodies at a distance, which is as good a definition of Eros as I know, the attraction of my body for the body of the earth and of the earth's flesh for my flesh. I am in a kind of ongoing erotic relation to the earth at every moment, as is each stone that I toss up into the air and that finds its way right back into contact with the ground. How different our experience would be if we recognized that gravity is Eros! It seems to me that, with a bit of attention, as well as a careful attentiveness to how we speak, we can begin to notice and make evident once again how wild and quite outrageous, still, is this interaction and reciprocity between our organism and a larger world that we did not create nor invent, the world that created us.

DC: For Abram, our belonging to the world, our reciprocity with it, is the foundation on which all our science and technology are built. It is also the foundation of ethics. In a recent essay called "Earth in Eclipse," he has a rather terrifying passage on the mass killings at Columbine

High School in Colorado. In this passage, he quotes a lifelong friend of one of the killers, who told a reporter, "What they did wasn't about anger and hate. It was about them living in the moment, like they were inside a video game." Abram goes on to speculate that perhaps these two young men who spent an extraordinary amount of time in virtual spaces had lost contact entirely with the sympathies of embodied existence; neither they nor the people they killed seemed altogether real. It is from the body, and the body of the earth, Abram says, that our sympathies come.

DA: Ethics is not, first and foremost, a set of rules or principles that we learn from a book or a teacher. It is, rather, a kind of sense in our bones. Ethics is the ability to move in this world without unnecessarily violating the ability of other bodies to move in their own way. It's a question of how to give space to others, how to restrain myself in cases where my exuberance or anger might impede the ongoing exuberance of another person. Ethics as how not to do violence, it seems to me, is something we learn, primarily, as bodily beings. We are living today in a technologically mediated world that cuts us off from our senses and has us living in a field of abstractions, almost oblivious to our bodily senses and to the sensuous earth around us, and therefore, I think, it's very hard to be ethical. It's very hard to know what it is to live in right relation with others, both with other people and with the other shapes and forms of our world. It's also very difficult for environmental activists like myself to mobilize people to act on behalf of another forest about to be clear-cut, another wetland about to be paved over and developed, or even against global climate change, because people don't really feel any deep affinity with the earthly world they inhabit. The body is our primary access to that world, and to the extent that we are living such disembodied lives, to the extent that we also are projecting the deepest truth of things into some dimension entirely transcendent to the sensuous, whether it be a heaven hidden behind the stars or a subatomic world hidden inside the nucleus of the atom, we're still not really living and participating with all of our intelligence here in this world. We don't feel any deep affinity with this world when we hear that other species are going extinct at unprecedented rates or even that the climate is changing rapidly. That's interesting, but it doesn't affect us very deeply because this isn't our real home. Our realer, truer home is somewhere else.

DC: David Abram has stressed that this feeling that real life is elsewhere has partly arisen from a scientific world view, but the problem for him is not science as such. He thinks that science, as a stance, has its place and its benefits. He objects only to the way in which science has overcome sensory experience and estranged us from the natural world. What he hopes for, finally, is a homecoming.

DA: I am in no way trying to disparage either science or technology. I am trying to argue that there is a world, a field of experience that we need to reclaim as our real home, as the primary dimension of human life and experience and relationship. That is the world of our direct corporeal, sensorial experience, the world of community, not of those with whom I like to dialogue on-line but of those whom I meet face to face, in the flesh, in my own neighbourhood. I want to give a certain primacy to place or to the sensuous, which is another way of saying the same thing, and to begin to practise a way of speaking, of languaging, that allows that primacy. This is not to say that we should not also pay attention to what's happening on the other side of the planet or that we should not engage with the Internet, but rather that we should recognize that there is one realm that is the soil, the ground within which all these other, more abstract dimensions remain rooted and from which they still tacitly or secretly are drawing their nourishment.

Science Manages the Sea

DEAN BAVINGTON

Management discourse seems to be continually changing and almost building on its past failures. This is what really surprises me about fisheries management now. Neither the failure of the cod fishery's management nor the ability of fisheries management to produce what it was designed to prevent has limited in any sense the proliferation of managerial designs and plans.

— Dean Bavington

On July 3, 1992, Canadian Fisheries Minister John Crosbie announced a moratorium on the fishing of northern cod. It was the largest single-day layoff in Canadian history — tens of thousands of people unemployed at a stroke. The ban was expected to last for two years, after which, it was hoped, the fishery would resume. But the cod have never recovered, and more than fifteen years later, the moratorium remains in effect.

The closing of the cod fishery concluded the strange history of the despoliation of what had once seemed an infinite abundance. When John Cabot sailed into the waters off Newfoundland at the end of the fifteenth century, he reported that the sea was so full of codfish that they actually slowed the progress of his ship. As late as the 1880s,

Thomas Henry Huxley, addressing a Fisheries Exhibition in London, claimed that the cod fishery was inexhaustible. There were, by this time, others who already doubted this opinion, but according to Huxley, "the multitude of these fishes is so inconceivably great" that no amount of fishing could seriously affect their numbers.

How this inconceivable multitude was hunted to the edge of extinction is a story in which science played an important part. From the outset, confidence in science underpinned the idea that the fishery could be comprehensively managed. This confidence was seriously undermined by the fishery's collapse, and today, consequently, the picture looks quite different. On the one hand, fisheries science has grown much more humble and tentative, much more aware of complexity and uncertainty. On the other hand, cod are currently being domesticated on fish farms, their life cycle manipulated "from egg to table," as the industry says — a project which has taken management to a level unprecedented in the old fishery.

Dean Bavington is a young Canadian scholar who has centred his studies on the way humans manage, or fail to manage, nature and on the different scientific models within which this management is carried out. His PhD research at Wilfrid Laurier University, completed in 2005, focused on the cod fishery. It's a subject in which he also has a personal interest, having lived and fished for cod as a boy in St. Anthony, at the tip of the Northern Peninsula of Newfoundland, where his parents worked at the Grenfell Mission Hospital. I talked to him about the role of science in the rise and fall of the cod fishery, and he began by telling me about how fishing was carried on in the centuries before it came under scientific management.

DEAN BAVINGTON

From year to year, fishermen would go out to a particular fishing ground, and, using the same technique in the exact same place, one year they'd catch a lot of fish and the next year they'd go back to the same place and catch very few. There were fluctuating landings from year to year, and in Newfoundland people adapted to these years of lean and years of plenty through occupational pluralism. You didn't just fish for cod. You would fish for cod for a little while and then you'd move on to other species. It was a seasonal activity. In other seasons you'd do activities on the land. This pattern of fishing was accommodated by an economic system called the mercantile system, or the merchant system: people would catch fish for their personal use, or subsistence, and then sell any surplus to the merchant or exchange fish with the merchant for products in the merchant's store. The number of these products, up until the 1850s, was very small. Catching the fish didn't require a lot of materials that people couldn't make for themselves.

The fishery for most of its history was primarily based on the hook-and-line method of fishing. It was pursued, both on the inshore and on the offshore banks, over a relatively short period of time when cod would migrate in following their main food source, which is the capelin. They would follow the capelin in from the offshore banks where they spawned, and fishermen would catch fish that were going after the capelin. They would catch capelin, cut them up for bait, and put them down on a hook, and when cod were hungry, they'd go for the baited hook.

Then, from the 1850s on, what starts to happen is that the mercantile system slowly starts to change into an industrial capitalist economy. It takes from about 1850 to 1950 for this shift to occur in the fishery. As this is occurring, new fishing technologies start to be introduced, and the amount of money needed to buy these goods increases. The merchants and bankers, the people who are investing in the fishery, start to problematize the fluctuations in landings because, with fluctuations in landings, you cannot guarantee a return on your investment. There begins to be a credit crisis in the outports. This is occurring not just in Newfoundland, but also in Western and Northern Europe. So governments, Norway being the first with respect to the cod fishery, start to say to natural historians, we need to understand what's causing these fluctuations so we can actually build a modern fishing industry that can depend on a steady and predictable flow of resources from nature.

DAVID CAYLEY
This demand that the fish should be brought under control in the wild was the beginning of modern management. Before this time, there had been management of the fish as a commodity — they had been

purchased, graded, and shipped into foreign markets — but there had been no thought of managing the wild fish.

DB: The nature, the behaviour of cod is not seen as something that is within the realm of control by humans. It's something that just is, that exists in nature as the providence of God. It's not something that humans can get involved with trying to affect. What I see as the major shift here — the management moment, you might say — is turning fluctuations, the ebb and flow of the codfish themselves, into a problem that is seen as solvable through some form of intervention, as opposed to something that is just given, in the nature of things, that we have to adapt to.

DC: The demand for steady and predictable fish landings was a product of many forces. The old mercantile economy was being gradually transformed into an industrial system. The cost of living was putting pressure on the fishermen to catch more fish. Bankers and investors wanted a guaranteed return. Politicians wanted to expand national economies. The agency that was expected to solve the problem was science.

DB: Governments start to ask natural historians, biologists, to come up with an explanation of what causes these fluctuations. The first part of the answer has to do with understanding fish as populations — single species populations that have their own dynamics. The codfish and our knowledge of the codfish was reframed. The focus shifted from the morphology of the individual fish to the aggregated scale of the population. By the 1930s, this population paradigm had become the standard way of understanding and studying fish.

DC: Population today is a completely taken-for-granted idea, so much so that it's hard to imagine how else one would think of fish or of people, but natural historians before the late nineteenth century had focused on what Bavington calls morphological characteristics, the shape or form of the creature in question. Their science consisted in describing a typical cod or a typical herring. Population was a very different kind of notion — abstract, mathematical, and managerial. Most important, it made cod conceptually manageable. It turned the multitudes of fishes into a comprehensible unit. Interestingly, though, Bavington says, there has never been consensus, even to this day, as to what actually constitutes a population of cod.

DB: There are debates between what are called the "lumpers" and the "splitters." The splitters find populations everywhere. They say there are lots of reproductively independent population units in the different bays and inlets. The lumpers say there's really only one reproductive unit. And, in fact, this debate between lumpers and splitters about the scale at which fish populations should be thought to exist continues up to the present. A lot of the distinction has to do with the stage of their life cycle at which fish populations are identified. When cod, for example, are spawning, they form large, concentrated spawning aggregations, largely offshore, on the offshore banks. Then they migrate back inshore. The idea was that, even though they may be spread out over large distances during the spring and summer months, these are not separate populations because they migrate back to spawn.

DC: Right. So doesn't that definitively settle the argument in favour of the lumpers?

DB: Well, you would think so, and for many years that's what's been assumed. But what we're starting to find now is that it's actually more complex than that. There are many different inshore populations as well — fish that do not migrate offshore to spawn but actually migrate between bays inshore and stay inshore to spawn. Studies that have been done post-collapse, post-1992, have shown the texture of the population as much more diverse and more like what the splitters had claimed.

DC: Cod populations, it has now been found, are more diverse and localized than the lumpers had assumed. This is something that many people fishing inshore had understood all along. This circumstance — that local people sometimes knew more than the scientists — will keep coming up as Bavington goes on with his story. Meanwhile, another concept had to be added to the population model before the cod fishery could be made fully manageable, and that was carrying capacity. With that concept in place, you could take the characteristics of the population — its fertility, natural mortality, and so forth — factor in the carrying capacity of the environment — available food, et cetera — and ascertain the surplus amount, the maximum sustainable yield, that could be taken from that population year after year.

DB: The maximum sustainable yield is a concept that supposedly lets you figure out how many fish you can take out of a population each year and have the overall numbers of fish bounce back within a year, due to the reproduction of the stock, to the level where it was the year before. This maximum sustainable yield idea emerges in an attempt to try and look at the dynamics internal to the population. It's framed in terms of the overall biomass or the weight of the stock, not in terms of individual fish that you're removing. Now, what's necessary for maximum sustainable yield to be meaningful or accurate is some knowledge of what the carrying capacity of the environment is. How much food is available is one limiting factor. The assumption which comes back to haunt fisheries management is that these underlying variables within the population can be held constant. There is an assumption that the natural mortality rate — the number of fish that are consumed by other fish or die naturally from year to year — and the reproductive rate of the population — that is, the number of eggs that are spawned each year by female cod and then added to the population in years to come — are both constant numbers. The idea of maximum sustainable yield is based on understanding nature as an equilibrium-based machine that is producing a surplus that you can skim off year after year. This idea develops from the 1930s, when the population paradigm gets established, and, by the 1950s and 1960s, it's fully established.

DC: The idea of maximum sustainable yield expressed the confidence of fisheries science that a comprehensive model could be made of cod and their environment. The capacity to exploit cod at this theoretical maximum depended on technology, perhaps the most important element of all in this story. From the later nineteenth century on, new tools and methods were introduced into the fishery at an accelerating rate, and one of the first was the cod jigger. The jigger was basically a weighted hook or hooks that could be jiggled up and down to catch some part of the fish's body. Today, it has an aura of quaintness and tradition, but at the time, it crossed a critical natural threshold. Previously, the fishery had ended when the fish stopped biting. Now the fish could be caught when they were lying satiated on the bottom, and the fishermen were quick to recognize what a momentous change this was.

DB: When the cod jigger was introduced, a lot of fishermen responded that it was inappropriate. They were concerned about the waste it caused, because many times they'd hook a fish in the side but it wouldn't actually come onto the boat. There was waste involved, and there were letters to the government at the time saying, you know, we need to ban this technology because it's wasteful of the fish. But that was not all. Fishermen also claimed that it would scare cod away, that they would not come back to the fishing grounds the next year. There was a lot of talk in these letters about the way in which the fish were perceiving this tool. There was also concern about what fishing this way did to the character of the person fishing. People argued that there was something unbecoming about this new technique, that it violated the proper relationship between the hunter and the hunted. With the introduction of this tool that, from today's perspective, is seen as traditional, you have resistance from many fishermen and a claim that it affects their own character as much as it affects the fish.

DC: According to Bavington, the majority of fishing people in Newfoundland actively resisted technologies that they felt threatened both the fish and their own integrity as hunters of fish. They petitioned the Queen, the governor, and the big buyers to ban destructive fishing gear. In doing so, they joined their voices to a chorus that dates all the way back to 1376, when English fishermen petitioned King Edward III to ban the practice of trawling, the dragging of nets along the bottom of the sea. The jigger was just one of many new tools that began to appear in the later years of the nineteenth century. There were new kinds of trawls, seines, and gill nets, new storage technologies, and the replacement of the single handline by longlines with many baited hooks. The crescendo reached its peak in the middle years of the twentieth century.

DB: The real change in the scale, as opposed to the character, of fishing comes with the arrival of the first bottom dragger. It happened in 1954, when a British dragger called the *Fairtry* arrived on the Grand Banks. From then until the late 1960s, you get the arrival of draggers from many different nations, from up to twenty different nations around the world, all fishing continuously — at all times of the year, in all kinds of weather. The peak landings of cod came in 1968. The numbers have never again been as high. Over eight hundred thousand tonnes of cod were caught, primarily by draggers.

DC: How does the dragger work?

DB: The otter trawl, which was the main form of dragging, works by dropping a net that's weighed down by huge iron wheels that roll over the bottom of the ocean. Cod is a ground fish that's down there on the bottom. You basically drop this thing down into an aggregation of fish that you've identified through a fish finder and run it through that school of fish — scooping them up — and then raise this net up to the surface. It's been referred to by some people as like clear-cutting the ocean floor, because it scoops up everything that happens to be down there. This form of fishing — not only the net itself and the hydraulic winches that are needed to haul it to the surface and to deploy it, but also the fish-finding technology and the steel-reinforced hulls — was an application of technologies developed during the war. Techniques used to find submarines were applied to the problem of locating and capturing fish in this post-war period. You have the arrival of the draggers, this landing of eight hundred thousand metric tonnes, and then a precipitous decline of catches after 1968.

DC: The bonanza of the 1960s created a crisis in the fishery. The solution, it was hoped, lay in extending Canada's jurisdiction to two hundred miles off shore, thereby excluding foreign fleets from all but the so-called nose and tail of the Grand Banks. In 1977, Canada proclaimed this new two-hundred-mile limit. But there was an interesting hitch. International law, by this time, held that a country that created such an Exclusive Economic Zone, as it is called, would have to give other countries access to any portion of its total allowable catch that it did not take for itself. This created a virtual duty to exploit the resource to the maximum. Fishing would have to be cut back for a few years to allow the stocks to recover, but after that, it was thought, an annual yield of between 300,000 and 500,000 metric tons should be available. And so the Canadian Department of Fisheries and Oceans began their management of their new realm.

DB: How were they going to manage this new territory? They were going to conduct annual survey trawls in the offshore and then extrapolate from these random trawls. In this way they would know not only how many fish were out there, but what the age distribution of the stock was — how many one-year-olds, how many two-year-olds, three-year-olds, and on up. From that they would then be able to estimate how many fish were being recruited into the stock each year and how many could be withdrawn. Once the commercial fishery started up again, they were going to combine that data with data from the commercial offshore fleet. This was called catch-per-unit-effort data, and it was based on the assumption that if it gets easier to catch a given amount of fish then the stock must be growing. So they combined these two numbers and got what was called the total allowable catch.

The problem was that after 1968 the inshore fishery really never recovered. They never did find it easier to catch fish, and that was despite the fact that people were buying more nets and more gear and spending more time at it. But this knowledge, this experience in the inshore, was not integrated into the management structure. What happened after 1977 was that, after three or four years, the survey trawls started to indicate a growing stock, and you got estimates of the number of fish that were going to be available in the early 1980s that were huge compared to what Canada had been landing traditionally. Because of these projections of the abundance that was going to be available, and because of the international agreement that what Canada didn't catch other countries had to be allowed to catch, there got to be a real emphasis on modernizing the fishery. The Canadian government helped to establish two nationalized fishing companies, National Sea Products and Fishery Products International, that owned offshore draggers and onshore processing plants. There was a really euphoric sense in this post-'77 period that we were going to get rid of this anarchy of the seas and have a rational fishery set up with our own Canadian draggers, as opposed to these foreign draggers.

DC: On this basis, a large-scale cod fishery resumed at the beginning of the 1980s. Inshore fishermen continued to have trouble finding fish, and their association took the government to court in an attempt to have draggers banned, a case that never made it past the provincial level. However, the offshore fishery appeared, for awhile, to rebound.

DB: The dragger fleet is finding it easier and easier to catch fish, and the survey trawls, in the early period of the eighties, are indicating the stock is growing. This happens right up to 1988-1989, when they start to see a discrepancy between what the commercial fleets are finding and the random survey trawls. The scientific survey trawls on the offshore are indicating that the stock is in trouble and previous estimates were probably too high. At the same time, the commercial fleet, on which this whole onshore processing industry is now reliant, is finding that their catch-per-unit-effort is actually going up. So you have these two sets of data saying different things. What the Department of Fisheries and Oceans decides to do is to average the two numbers. Two years later, the dragger fleet goes out to try and find their allocated catch and can't find any fish. John Crosbie, the fisheries minister, just back from the Rio Earth Summit, where he was advocating a conservation and sustainable development approach to the marine environment, comes home to an industry that can't find any fish. There's a crisis. The fishery is shut down on July 3, 1992. The assumption is the same as in 1977 — we'll stop the fishery for a few years and the stock will recover. But the cod don't recover after two years. They don't recover after five years. And they don't recover after a decade.

DC: The collapse of the cod fishery raised many questions about the scientific assumptions on which it had been based. Modelling fish populations and their marine environments as simple, machine-like systems, it seemed, had simply left too much out. One thing that had been overlooked in the 1980s, for example, was the over-all age structure of cod populations.

DB: It's now understood that there were permanent effects from the killer spike, as it's called — the huge landings up to 1968 and then the drop-off. It turns out that what you need for a population that can continue to reproduce itself is a lot of large, old female fish. The larger the female is, the more eggs she produces and the more survivorship occurs from those eggs. After 1968, there were basically no old fish left in the population — male or female. They'd all been removed. The assumption of the fisheries models was that as long as fish were, in fisheries lingo, allowed to "recruit to the population," as long as fish were allowed to get up to reproductive age, which was age six or age seven, as long as they were able to get up to that age and spawn once, then they could be removed. They'd done their job in terms of adding to the next generation and maintaining the maximum sustainable yield. But what is coming out now is that that's not the case, and that really everything that happened after 1968 was insufficient for the stock to actually recover. The damage had been done.

DC: One of the interesting things about this belated scientific discovery was that fishing people had worried about damage to what they called "mother fish" going all the way back to the 1850s. In the period leading up to the moratorium, according to Bavington's research, it was the inshore fishermen who were the first to recognize what was happening. They knew, first of all, that they were having trouble finding fish. They knew that the fish they were catching were smaller for their age than formerly — a point I'll return to — and they knew that the offshore draggers were catching females during spawning.

DB: When cod spawn, they aggregate, and they're actually easier to catch because they're not randomly distributed all over the fishing grounds. There were stories coming back of draggers dragging up cod that had spawn in them. It didn't make any sense to allow this, so they advocated banning the offshore fishery during the pre-spawning and post-spawning period. It seemed like common sense. Why wouldn't the regulators adopt this? Well, it turned out that the offshore fishery *was* controlled in terms of when fishing could occur, but it was based on an incomplete understanding of cod spawning — fishing was allowed during a period of time when the cod were thought to have already spawned but hadn't. The timing of spawning, it turns out, shifts from year to year. They don't spawn at the same time every year. So the regulations were not mapping on to the biology of the fish. This fact seemed to be recognized more and understood more by the inshore fishery than by the offshore.

DC: Inshore fishers, in Bavington's view, generally recognized more variety amongst cod populations than was allowed for in the scientific models that authorized the offshore fishery. This was true not just of the timing of spawning, but also of the makeup of the species. The allowable-catch, sustained-yield model basically treated the cod as a single inventory. Local people in Newfoundland recognized various types of cod.

DB: In the vernacular, there are six or seven different ways in which cod are talked about. The scientific name, *Gadus marhua*, refers to a single species, but in the inshore they have six or seven different names for cod. They refer to cod from different bays, and they're based on qualities that they can distinguish, like the colour of the fish, the taste of the fish, the timing of when the fish come in, and so on. The inshore fishermen did not understand northern cod as one unit stock, one large population. They saw that there were some cod that stayed inshore and some that seemed to migrate between bays. It was a much more fine-grained understanding of the makeup of the fish than occurred in the offshore.

DC: How did the scientists relate to the knowledge that the fishers had?

DB: Well, the way in which they related was to call this knowledge of the inshore fishermen anecdotal. These were stories that couldn't be validated one way or the other. There were no numbers behind them, and they couldn't really be trusted. They didn't give an overall picture of what was going on. Knowledge from the inshore fishery was not only disparaged as anecdotal, it was also opaque from the point of view of the managers. The only information that was considered reliable was the numbers from the scientific survey trawls, which were occurring offshore, and they were using the same technology, the dragger, as the commercial fleets were using to produce their numbers. So, in one sense, the knowledge was disparaged, but, in another sense, it was a dark space. It wasn't knowledge at all.

DC: Before 1992, the knowledge of the inshore fishers was not generally given the same standing as scientific knowledge. The word "anecdotal" sums up this disregard. These were mere stories. Scientific knowledge was complete, reliable, disinterested. In retrospect, though, Bavington says, it's clear that these were two distinct and different styles of knowing.

DB: The inshore fishery, framed as anecdotal, is very much tied to place, to particular places, and also tied to the senses — what you can see, what you can taste, what you experience out on the water when you're hauling your nets or fishing with hook and line. Against this are the quantitative representations of the fishery that occur in graphic form, that occur on computers, that have to do with seeing trends over longer periods of time, and that are integrated into management. And, really, from the very beginning, there was no distinction between pure and applied science in fisheries. It's always an applied science. It's funded from the beginning to solve a particular problem, which is the fluctuation in landings. Science is supposed to be able to go out and make the sea legible from afar. How many fish were available out there? This number becomes the focus of everyone involved in managing the fishery.

There are several problems with this. First of all, in order to obtain that number, the scientists needed to fix a lot of the variables — to assume that they were the same from year to year. Without this assumption of equilibrium, the model would become too complex to calculate. Because there was no way to actually go out year after year and find out what each of the different sub-populations' birth rates, reproductive rates, were, they relied on averaging. The average fish starts to emerge. The average female fish will produce an average number of eggs that will survive an average amount of time, et cetera. These averaging assumptions are, in a way, a conceptual domestication of nature. What appears on the surface as an unmanageable, wild area gets represented as an equilibrium machine, as something that, at least in the mind, comes to be seen as graspable and controllable. The amount of fishing is the only thing that is not a constant, and, by limiting fishing, by changing total allowable catch levels, we can regulate the fishery. This very simplified model of wild nature allows a certain type of hubris, I think, to develop. We control fishing effort, hold all the other variables constant, and are then able to project out into the future what will happen to the stock, and, based on that, how to develop Newfoundland and Labrador. It's a lot different from the coping and adapting to the dynamics of cod that occurred before this period. That's what the inshore fishermen seemed to be always pointing to: it's not something that we can control, these representations of the fish are not matching up with what we're experiencing in our individual bays and inlets.

DC: Fisheries scientists, Bavington says, worked from the beginning in contexts where powerful interests impinged on what they did. Governments, eager to maintain employment, and processors, eager to keep their plants working at full capacity, were also part of the annual closed-door meetings at which the allowable catch was set. Even so, unwarranted confidence in the idea that a sustainable maximum could be calculated and relied on certainly contributed to the downfall of the cod fishery. The closure of the fishery in 1992, consequently, led to a rethinking of fisheries science and to the adoption of what Bavington calls an ecosystem view of cod and their environment.

DB: The ecosystem view, the ecosystem understanding of cod recognizes that all the variables that were held constant are actually in flux and changing, that interactions between species — predator-prey dynamics — are important, and that the act of fishing, once assumed to just be skimming off surplus production, actually changes not only the environment — disturbs habitat — but also changes the genetic makeup of the population itself. If you think about it, what fishing ends up doing is taking out all the large fish. From the perspective of an individual fish, if you're to survive this fishery, you don't want to grow very big. You want to be small enough to get through the mesh of the net when the net's coming. You want to be able to pass on your genes before you're scooped up out of the water. So what they have found is that heavily fished populations like the northern cod tend to start maturing, that is, reproducing, at earlier and earlier ages and at smaller and smaller sizes. This is very new science. This is within the last few years. But the interesting thing for me is that we ever saw the relationship between fishing and the fish as one in which we could just reach into a population and skim off a certain biomass of fish without affecting the very nature of these fish. With the rise of this new science, the focus has shifted to relationships and interactions, as opposed to the idea of fish as objects or that fish are produced as if in a machine. The environment in which cod are embedded is understood to be dynamic and in flux. The stance that tends to come from this is that scientists are saying, we can't give you that maximum sustainable yield number. The error range in what we're going to come up with is going to be so wide that it's not going to provide managers with something that's usable. It's as if the more we learn about codfish, the less useful that knowledge is to management. A more complicated story emerges, and what fisheries ecology is now generally putting forward is that there's no way in which we can actually build a rational fishery based on the knowledge that we have. This leads to a stance of coping and adaptation, which is very different from the science from the pre-1992 period, when it was much more confident. I think there is a more humble view vis-à-vis wild fish.

DC: Wild is an important qualifier here because since 1992 there has also been a move to domesticate cod and raise them on fish farms. This move, in Bavington's opinion, stands in dramatic contrast to the new humility he sees in the study of wild fish. Some small-scale forms of aquaculture have been around for thousands of years, but the attempt to domesticate species like salmon and cod, he says, requires a more comprehensive management of nature than was ever attempted in the old fishery. One of the biggest problems arises from the fact that cod, like salmon, are carnivores. They eat other fish.

DB: The argument that we often hear from the aquaculture industry is that the expansion of aquaculture will actually help wild stocks, wild fish stocks, because, well, you're not out catching them anymore. You're growing them. On the surface, that appears to be correct. But if you're growing a carnivorous fish, you have to feed it other fish. To grow a pound of salmon, you have to feed it three pounds of wild fish, and this wild fish most often doesn't come from the area where you're actually growing the fish — it comes from stocks that are caught primarily in southern waters. Off Peru, you have a very large anchovy fishery in which the anchovies are caught and turned into fishmeal, which is like a fish flour, basically. This meal is combined with fish oils and other ingredients to feed to carnivorous fish like salmon. So, over the long term, fish farming doesn't take any pressure off wild stocks, if what you're growing are these carnivores, that have been called tigers of the sea. When you're eating a carnivorous fish, it's as if you're eating a tiger that has had to be fed cattle.

DC: The fact that farming carnivorous fish destroys more food than it creates is just one of the problems that Bavington, along with many others, connects with this new technology. Fish farming is developing with astonishing speed when compared to the pace at which plants and animals were domesticated on the land. Fish farms spread disease to wild fish and attract predators — seals, otters, birds — that have to be killed or kept away. Escapees interbreed with wild stocks with unpredictable consequences. And in Newfoundland, Bavington says, there's an additional problem. The only bays with the ice-free conditions cod aquaculture requires — Placentia and Fortune Bays on the southwest coast — are the very places where the most vibrant populations of wild cod still survive. Fish farming, moreover, seems to be changing the very idea of fishing, so that even fishing in the wild is now described in domestic terms.

DB: The way in which fishermen are talked about these days is as fish harvesters — not as fishermen or as fishers. The wild fishery that still exists in Newfoundland has moved on to other species — mainly crab and shrimp, and lobster — and those fishermen who want to stay in the industry now have to become what's called professional fish harvesters. They have to actually go to college and get certification in order to have the right to obtain a licence and to obtain a quota for these other species.

There is this dual domestication that, I think, is changing the identity of the fisher to that of a harvester, that of someone who individually owns a quota for a particular species. In the pre-1992 period, it was mainly the state, the Canadian government, that was seen to be the owner of fish populations. As well, you have what's been called the criminalization of subsistence fishing. It's now illegal to catch a codfish in Newfoundland for your own consumption. The type of fishing that I was doing off the point in St. Anthony when I was growing up I now could be arrested for. And people *have* been arrested and charged and fined and have had their boat, if they're out in a boat, or their car or truck, whatever they used to get to the place where they're going fishing, all confiscated. What I find very troubling is that it's now illegal to catch a fish to eat, but other industries that are targeting commercial species of fish are allowed cod by-catch. When you drop the dragger net, you don't only catch one species, you often catch a mixture of different types of species. People who are targeting species like shrimp are allowed to have a certain percentage of the shrimp actually be cod. And that's what cod fishing now has become — incidental by-catch.

DC: Dean Bavington began his study of the cod fishery out of an interest in the different stances humans have adopted towards nature. For centuries, Newfoundland fishermen hunted cod only in the season when they were running and biting at bait. They accepted fluctuations in their catches as part of nature's uneven providence, adapted themselves to the habits of the fish, and preserved a sense of honour in relation to their prey. Then came the commercial, technological, and scientific revolutions that resulted in the ruin of the species once known as king cod within a space of about 150 years. A recent report by a respected research group at Dalhousie University predicts that, at present rates of exploitation, all of the world's commercial fisheries will be in the same state by the year 2050. This history, and this prospect, have brought Bavington to the conclusion that management has failed and that it may be time to go back to a style of fishing that preserved the existence of the fish and the integrity of the fisher. But so far, he says, he sees no abatement in the desire to manage nature.

DB: Management discourse seems to be continually changing and almost building on its past failures. This is what really surprises me about fisheries management now. Neither the failure of the cod fishery's management nor the ability of fisheries management to produce what it was designed to prevent has limited in any sense the proliferation of managerial designs and plans. Most of the discussions that are occurring within the fishery are along the lines of, my managerial method is better than your method. Fisheries management has also taken on the idea that it's no longer the fish that should be the target of control but the fishers themselves. The technology still gets left out, but the fisher person is very much the target of management.

For me, I think it's much more interesting, at this point, to start to take seriously the words of the inshore fishermen, who say, basically, that we have the technology now to destroy all the fish and eliminate our way of life. What we need to do is figure out how to get back to baited hook and line fishing and enable its continuation, as opposed to fish farming or fish harvesting. Something else that I think is really interesting, too, is that, pre-1992, as we've been saying, the inshore fishermen were shut out. Back then they were saying, there's no fish, manage the technology in the offshore, shut the fishery down, there's something wrong, we're not catching any fish, while the scientists were saying, there are lots of fish. Now what's happened is that the scientists are saying, there's no fish left. The cod are on the verge of extinction. They should be declared endangered. Inshore fishermen are saying, no, there are fish left. There's not enough fish for a commercial fishery, but there are enough fish in the inshore for us to be allowed to catch some of them to eat for ourselves. And different bays have different levels of fish. This is an interesting thing. The scientists are still in disagreement with the fishermen, but the fishermen are now saying there are fish, while science is saying there aren't. It seems to me that the only way to get into a conversation that will move beyond these polar opposite positions is to start to take the role of fishing technology seriously. I don't think that we can manage our way out of the problem as long as technology is still seen as this thing that, if managed correctly, can be sustainable. The question of technology is not just about how we do things but about how we want to live. Fishing is a way of life and not just a means for some other end. But since Newfoundland joined Canada in 1949, that is how the fishery has been seen — as something that we could use as a stepladder to a higher stage of development and then get rid of, leave it behind, instead of thinking of it as something that still has some role in present ways of living.

What Genes Say

BARBARA DUDEN

Genes had the capacity to redefine the subject, the person, in its entirety, in every cell. Genetics belongs to the history of the disembodiment, the disincarnation, of your own orientation — the answer to the question, who am I? Genes have become the answer to, who am I?

— Barbara Duden

When Danish botanist Wilhelm Johannsen gave the infant science of genetics its central concept in the early years of the twentieth century, he described his coinage, the "gene," as a "very applicable little word," and so it has turned out. Once a purely scientific and technical term, it has now spread into common, daily use. People speak familiarly of "my genes" and "your genes," newspapers report on the latest "gene find," and an American company now offers anybody with a thousand dollars and a saliva sample the chance to have their genome mapped. Under the slogan "Genetics Just Got Personal," the company's website invites browsers to find out "what . . . your genes say about you."

But what happens when a scientific term migrates from the laboratory to the street in this way? What does the word "gene" signify in everyday speech? German scholar Barbara Duden, of the University of Hanover, has been pondering this "pop gene," as she and her collaborator Silya Samerski call the gene in popular culture. Most talk about genetics is concerned with what genes *do*, with whether this or that gene gives you blue eyes or a good golf swing or an increased risk of heart disease. Duden is concerned with what genes *say*, with what they tell people about who and what they are.

Barbara Duden is a historian and a pioneer in the field she calls "body history," the history of how people have actually experienced embodied existence. When she began writing history in the 1970s, her fellow historians still tended to diagnose the illnesses of people in the past in modern scientific terms. Her innovation was to take her historical subjects at their own word. In her first book, *The Woman Beneath the Skin*, she drew on the case records of an eighteenth-century German physician to bring to light the vivid ways in which women had felt and imagined their pains and pleasures in an age before objective measures of the body's condition were available. Her second book, *Disembodying Women*, looked at the contemporary female body, mapped, monitored, and measured, as the eighteenth-century body had never been, but also objectified, deprived of its story and its voice. Her current work on genes carries on with this theme of disembodiment.

We spoke about this work in an interview we recorded at her home in Bremen. The popular image of the gene, she told me, sums up a deep change in the practice of medicine that she thinks began in the 1970s. She gives the example of hormone replacement therapy, the artificial supplementation of hormone levels in women around the time of menopause.

BARBARA DUDEN

In hormonal replacement therapy, we could recognize that there was a totally new logic that came about because hormonal replacement therapy is not a therapy in the sense in which there is someone sick, and then you do a therapy. When a physician offers hormonal replacement, he doesn't give a diagnosis because the woman isn't sick. She's just getting older. But he will tell her, when you are thirty years older and you fall down the stairs, you will break your bones. It's better to do something now, so when you fall down, your bones won't be as brittle. There was this propagation of risk consciousness and risk management. We could see this not only in hormonal replacement therapy, but in many practices. Physicians had changed from dealing with the sick to managing risky selves, to managing people who are perfectly healthy but are afraid something might happen in the future. They are being told in what ways they can manage their present being better, so that, in the future, they will have better chances. This transformation was a transformation of the flesh from stuff that, in the ordinary, is very well and, only in certain instances, when you are sick, requires a physician, into something that now needs constant rectification, investigation, monitoring — something that might already be sick, even if you don't feel it yet. It seemed to us that genes were that Trojan horse through which risk consciousness would be implanted, not only into your interior, but into every cell, so that in every cell, in the very substance of your being, the gene is the incarnation of risk.

DAVID CAYLEY

Duden decided to test her surmise that genes constitute the Trojan horse by which risk awareness is implanted in people's bodies. She and her colleague Silya Samerski began to look into the ways in which genes were infiltrating everyday life. As part of this research, they conducted interviews in a German village, a place at the dead end of a rural road but still near a university town with a celebrated institute of molecular biology. There they found that everyone spoke freely and confidently about genes.

BD: The first thing that struck us was that this term had been popularized so much that, in this small village in the south of Germany, everyone would very vividly talk about genes. The word has really entered into ordinary conversations involving the woman in the bakery, the woman who is cutting hair, the minister — we spoke with the minister of the village. He would pontificate that God made the genes. Then the woman who is cutting hair would talk about the genes and cancer, and the local physician would talk about how drunkenness is really genetically determined. There were all these stories — wild, wild stories. First of all, we were amazed at the astonishing richness of this popular gene talk.

DC: The richness of popular gene talk results, in Duden's view, from the fact that the term has no precise denotation for the people using it. No one could satisfactorily define it, and many had fanciful ideas about the form and location of genes. This lack of definition is one of the keys to the word's remarkable power. It can signify so many different things, beginning with heredity.

BD: We spoke with a woman who sold sausages because we found her explaining genetics over the counter to a client. She said, "The chromosomes are as long as this," and she demonstrated with her finger the length of the chromosomes. She had had genetic counselling, and we were interested in what was left with her. What we found out is, first, that genes stand in for how you are in your very flesh, as the daughter of this mother and grandmother and so on. So, she would say, "I can see in my children the genes of their father — how, when they sit there, they sit like him. One daughter is unruly. He also was, in his childhood, very unruly. So this is her genes. We can see it." It is like a synonym for saying, this is how she is because she's the daughter or the granddaughter of this person. Genes stand in for something that is given, handed over in inheritance.

DC: The second set of meanings that people connected with genes had to do with risk, Duden found. Genes point to troubles which may break out in the future, as another woman from the village explained.

BD: She said, "But maybe the gene is like a capsule. It sits there, and then suddenly it opens, and you don't know when it will open. Maybe you ran too fast, or you did something wrong, and then suddenly it opens, and then the sickness comes." She speaks about genes in the image of a latent threat, something that's already there but only latently, something wrong that will show up later.

DC: Genes can refer to risk, they can refer to heredity, and they can refer finally to the possibility of bringing life under a new kind of management. A number of Duden's interviewees evoked a utopia in which genetic manipulation would erase every inconvenience of the human condition.

BD: I take the example of the woman who was selling sausages in the village. She would talk at length about how ultimately, with genes, you could redesign human beings so that everyone has blond hair or whatever — all these bio-fantasies about what could be done. Also, what was very important was that, when they were talking about the possibility of manipulation, there were all these ideas about how, in grave sickness, genetic research ultimately would be the answer. The gene is tied to hopes and promises of a future in which there is no sickness, no old age, you don't have to die, and so on. People imagine the possibility that the human being, in its flesh, could be manipulated, remade, and redesigned in its entirety.

DC: Duden thinks that people understand genes in confusing and contradictory ways. The word for her operates in two very different registers. Genes signify something intimate, tangible, and personal — the eyes I share with my mother, the walk that makes people mistake me for my father at a distance — and at the same time, they point to an intangible and impersonal realm of risk, dependency on experts, and the possibility that everything might be about to change. The effect of these discordant registers is to throw people off balance.

BD: With genes, anything may be possible, anything could happen, and because this word pertains to so many contradictory dimensions, I think people lose the ground on which they can orient themselves. They really become dependent on professional counselling. "Gene" is a word that destroys the very possibility of saying, I am well, I feel well. We'll see what happens in the future. I'm in the "now." I know from where I come, and I act accordingly. With genes, there may be already a defect in yourself that will show up later, even if you don't feel it. You cannot really influence it, so it's destined, while at the same time, you've got to continually investigate it to see if you can manage your chances better. On top of that, it is something that might be able to change the face of the world in its entirety. It is related to ultimate promises of the annihilation of sickness, getting old, and so on, which makes it hard to say no to it. So it's an extremely powerful term, and it's extremely powerful because of these contradictory modalities that come together in a word which, at the same time, says something about who you are and who the next one is. We thought that the most important effect of genes is not in the sphere of science, but in the symbolic fallout of gene talk that brings about "pop genes."

DC: The pop gene, as Duden sees it, derives its power from its ability to bind together so many different ways of talking into one word. It embodies and reconciles contradictions, holding together what she calls different spheres of existence.

BD: The word covers spheres that ordinarily cannot be covered in one word. It's a word for where you come from — something very personal — but it's also a word that's apt for, we could say, the management of hope, a word that elicits possibilities of what could be done to the human condition. This possibility that the human condition could be unhinged, could be, in a deep sense, changed, is, of course, tied to experts and the world of laboratory research.

DC: The pop gene builds a bridge between personal life and the scientific laboratory, Duden found. The scientific gene, by definition, is something that can exist only in a fully equipped laboratory. Even there, the boundaries of the object will change with the ebb and flow of scientific knowledge. Indeed, some scientists now argue that the advance of genetics has turned the very idea of genes into a misleading oversimplification of the complex mechanisms of heredity and development. The pop gene bypasses these difficulties. The peculiar and exacting conditions underlying the production of scientific facts are erased, and the gene becomes an everyday object, as real as any other object of sense. The gene is naturalized, Duden says, made into an intimate part of one's own nature. Her concern about this is that it conflates spheres of life that are of radically different kinds. Again and again in her work, she has drawn attention to the danger she sees in this merging of the personal and the technical. For example, in her book *Disembodying Women*, she talks about ultrasound scanning of the pregnant womb and argues that a crucial distinction must be preserved between the expected child and the ghostly apparition on the ultrasound screen.

BD: The ultrasound screen is a technical device for the discovery of abnormalities. The physician looks and screens and checks whether the unborn has any fault or is in accordance with the average, but the woman sees the child she's going to love. You have a merging between a monitoring or screening gaze at an object that is to be evaluated according to standards and the emotion you feel for the next person you are going to love. Technical supervision merges with the intimate and delicate and deep perception of love that you feel when you see your dearest to come. You see the beloved in a framework in which its quality is being evaluated, and if it's not up to standard, you may be asked to make the decision to get rid of it.

DC: Duden wants to maintain a firm distinction between the realm of techno-science and the personal realm, between the objectified fetus picked out by the ultrasound scan and the beloved to come. She sees the gene, in an analogous way, as belonging to an entirely different order from the lived body. Let this distinction break down, implant the gene in the living flesh, and human personality itself changes.

BD: Genes have the capacity to redefine the subject, the person, in its entirety, in every cell. I think, when we look back, we can say that "the gene" was important in redefining the person, in its past, present, and orientation towards the future — in its very flesh. The importance of genetics is in the history of the disembodiment, the disincarnation of your own orientation — the answer to the question, who am I? Genes have become the answer to the question, who am I?, and the answer they give to the question, who am I in my flesh? is deeply contradictory and, especially, disorienting. You lose your common sense. The moment you believe in genes, you incarnate a destiny that's already given, while, at the same time, you are called on to practise good risk management. The belief entices people to experts, even when the expert ultimately cannot help you.

DC: If you believe in genes, you lose your common sense. Duden's statement, as I understand it, says nothing whatsoever about the gene as a scientific construct. Scientific constructs, in any case, do not invite our belief. They require the kind of careful and qualified assent which is only possible for those who actually know something about the complex reality that a term like "gene" summarizes and simplifies. What Duden is arguing is that such scientific terms have a mystifying and disorienting effect when they stray into what she calls the "prose of everyday speech." There, she says, the gene simply has no place.

BD: The word is an alien in our prose. It is something that unhooks one's perception from one's biology in the traditional sense of the way in which I can make meaning of my own life. The difficulty is that I cannot incarnate a scientific gene. The scientific gene pertains to statistical populations, but people orient in the flesh. Gene talk implants genes as causes. The pop gene becomes the cause of that which is and also of something that might happen in the future. When you ascribe that to yourself, you involve yourself in contradictory orientations to your own existence. You incarnate or incorporate a cause of your being that is, on the one side, destined, but that also has the aspect of conditionality. It may be dangerous. It may be fearful. It's there, but you can't feel it. It may show up in the future, so it takes the bottom out of your experience. This cannot be mended by demanding genetic literacy, because genetic literacy will only estrange people further from the only way, I think, that you can really orient, which is to trust your own sense, to trust the experience that you have acquired in your life as a person in flesh and blood, where you know something. When you speak about genetic knowledge, it makes you totally dependent on whatever the experts tell you. It cannot be reached from what you actually know.

DC: What, for you, is the proper relationship between everyday living and scientific knowledge?

BD: I think, the older I get, that it's not good for very much. It's not very helpful. It's really not very helpful. We can analyze it, as social scientists, and show how useful gene talk has been as a resource for all types of things: to manage hope and make people believe that ultimately medicine will be able to annihilate sickness and so on. But in your daily experience, it doesn't help you. In your daily experience, it's very important to stick to your senses, or to whatever leftovers of the senses remain. It's a problem of the academics that they disincarnated or disembodied their orientation, so they take the world as the outcome of experiments in the laboratory. But the beauty of the real world, the "I" and the "you," where we are real, is rooted in the senses. To be alive, you must sit in your flesh. That's the most important.

Losing Sight of People

SILYA SAMERSKI

The first step in statistics is to lose sight of the concrete person. It's only about numbers. It's not about people made of flesh and blood.
— Silya Samerski

Use of the word "gene" goes back to the beginning of the last century, when Wilhelm Johannsen derived the term from the name William Bateson had just bestowed on the projected new science of genetics. But it was not until DNA was modelled in the 1950s that the term was given a precise form and location: a gene was a stretch of DNA which comprised the blueprint, so to say, for some part or propensity of the organism. Since then, as the science of genetics has grown in subtlety and refinement, it has largely dispensed with simple, mechanical, deterministic images of how genes work. But at the same time, genes have entered popular culture as the quasi-mythic power that Barbara Duden has just described: the pop gene that haunts everyday conversation in a German village.

Pop genes are put into circulation in various ways. The media need a handle on genetics, and genetics itself needs an icon of its power and promise as a science. Medicine, too, plays its part, and its role has been the particular study of Silya Samerski, Barbara Duden's collaborator. Samerski sat in on the genetic counselling sessions that doctors give to pregnant women in Germany and later published her reflections on what the women made of what they were told in a so-far untranslated book entitled *The Mathematization of Hope*. We spoke in Bremen, where she lives, and she told me that her interest in genetics traces back to the time in the late 1980s when she went to the University of Tübingen.

SILYA SAMERSKI

I started studying biology in Tübingen, and, like quite a few of my colleagues, I was interested in nature. But we soon found out that biology is not about nature at all. That was pretty disappointing, actually. Then the university started to change the whole structure of the field of biology. It cut down the classical disciplines like botany, zoology, and so on, and started to build a centre for gene technology. All these changes that have taken place in biology in the last twenty or twenty-five years were just beginning when I was there, and quite a few students opposed these changes because we were very critical about gene technology. First of all, we thought of it as a risk technology. It is risky. Then, it's not about understanding nature. It's a technology. It's about changing something. It's about making money. I think that was how it started.

DAVID CAYLEY

Samerski and her fellow students were unsuccessful in their efforts to convince the University of Tübingen to turn away from gene technologies, but in the course of her critical work on the issue, she concluded that she needed to know more about what she was dealing with.

SS: I thought maybe I should take the chance and study human genetics. Since I was already doing some political work on genetics, I thought it would be a good idea not only to criticize it, but also to go inside of it and really study it and see how these institutions work — really get into it. So I did that. I chose human genetics as my main discipline. Thus I had one leg outside of the laboratory, discussing critically with my friends the political and social consequences of genetics, and one leg inside the laboratory, watching everything or even doing genetics myself. That was very fruitful, I think.

DC: Silya Samerski earned her diploma in human genetics at Tübingen in 1996. During the course of her studies there, she became aware of the phenomenon that would eventually become the subject of her doctoral dissertation at the University of Bremen: genetic counselling.

SS: When I was working in the laboratory in human genetics, I always saw these people coming in, parents with children or couples or women. They were waiting for the genetic counsellor in front of my laboratory, and I saw them. Most of the time, they were very nervous, and I wondered what was going on there. Of course, since I was qualifying in human genetics, I was also taking seminars on genetic counselling and prenatal diagnostics, so at first I asked them if I could take part in one of these sessions, and they said, yes, of course, no problem. I just listened to those sessions, and they really struck me. I was very surprised at what was going on there.

DC: A number of things surprised Silya Samerski about these encounters between pregnant women and doctors with special training in counselling and genetics. One was the sheer amount of information that was gathered from the women, beginning with the counsellor taking a family history.

SS: They draw — which is a big part of the genetic counselling — a pedigree, and it's really called "pedigree," Stammbaum. They go through all the relatives: aunts, uncles, grandparents, and so on and so on. As you can imagine, in all families, there are illnesses, someone died early, that's just how life is. But everything that has happened in the family in the genetic counselling session is turned into a risk. Let's say there was a child that died very early of an unknown cause — of course, most of the time, it's of unknown causes, maybe it was shortly after the war — so they draw a circle around this case and say, maybe you better check again why this one died. It might be genetic. It could be genetic. We can't rule it out. Also, many counsellors made it an issue when someone had a heart attack. Everything that has ever happened and is considered somehow genetic is turned into a risk that the client should guard against, either a risk for herself or for her child.

DC: This first stage of genetic counselling creates a risk profile. The counsellor then teaches the client about risk and about how to decide whether further genetic screening is warranted. Here Samerski observed what she came to think of as a crucial ambiguity. The client could hardly help but think of the risks being ascribed to her as something personal, as statements about her, whereas, in fact, they only describe statistical regularities in large populations.

SS: If a doctor tells you you have a risk of one out of twenty of developing this cancer, it just means that he has put you into a statistical population in which one out of twenty is considered to develop that cancer, but nobody ever knows if you will develop it or not. You don't know whether you have anything in common with those other twenty people except maybe this one feature that made him put you into that statistical population. The first step in statistics is to lose sight of the concrete person. It's only about numbers. It's not about people made of flesh and blood.

DC: Risks are impersonal. They describe not what will happen to me, but what might happen to someone who resembles me in some respect. Nevertheless, risks are what a pregnancy is made of for a woman in genetic counselling.

SS: The simple fact that she's pregnant creates what's called "basic risk." Once someone told me, it's like when you enter a taxi. They start with five dollars. If you become pregnant, you start with a risk of three to five percent that the child might not be normal. Then those counsellors make a list of what could be wrong with the child, so they tell the woman, who somehow should be in good hope, your child may have a congenital heart defect. It could also have some other genetic disorders. It could have a harelip. That's, so to say, the general information she should receive to be a responsible pregnant woman. That's what every woman should know nowadays: what might be wrong with your child, the probabilities of all these disorders and so on.

DC: A pregnant woman was once said to be "in good hope" or just "expecting." Today, it seems irresponsible to leave any possibility unpredicted. Samerski herself has two young children, and she contrasts her own case with her mother's.

SS: In Germany, we have this mother passport. You probably don't have that in other countries.

DC: Not to my knowledge.

SS: It's really called *Mutterpass*, Mother Passport. It was introduced in '68, I think. I looked at my mother's Mother Passport when she was pregnant with me, and it had maybe four pages. All that was in there was her weight, when she saw the doctor, which was maybe four or five times during her pregnancy, when she felt quickening for the first time — things like that — very, very little. If you looked at my Mother Passport, you wouldn't believe it: a lot of curves and so many entries and fifty-two risk factors that I could have. It's a huge change within the last thirty-five years. When my mother was pregnant with me, there was no such thing as prenatal diagnostics. There was no ultrasound. There was no amniocentesis, no genetic screening. There was no screening at all. She was just pregnant, and the doctor maybe took some blood and took her weight, and that's all.

DC: Prenatal diagnosis is the ultimate point of genetic counselling. Does the risk profile indicate that further testing is in order? If it is, this will involve some combination of ultrasound inspection of the fetus, blood testing, and perhaps amniocentesis, which allows the culturing of fetal cells that have been withdrawn from the amniotic sac with a large needle. It was up to the women, after counselling, to decide whether to take this next step, and, for many of them, Samerski found, it was a puzzling question.

SS: I saw many women who were at a loss after the counselling session because it's a very abstract decision. It's somehow like a managerial decision or like a decision you make about the stock market. You have various risks, and you try to increase your chances. But in the case of a woman who's expecting a baby, this kind of mentality doesn't make any sense. First of all, she's expecting one baby. What does it mean to her, what can it mean to her, if she hears statements like, you have a risk of one out of 180 that your child might have Down syndrome? What can it mean? She'll have one baby, not two hundred. She wants to know if this baby that she's carrying will have Down syndrome, or anything else. It doesn't say anything meaningful to her. It can't say anything meaningful to her. I remember one session where the woman said, "What does it mean?" The counsellor said, "That's up to you. You have to decide. Some people consider that low. Some people consider that high. If you consider this a high risk, you can undergo amniocentesis." It shows the nonsense of this whole procedure because even the experts can't really tell women or their clients what that should mean. They're also at a loss.

DC: The decision as to whether to undergo further testing is often imponderable, and even when these tests are done, the results may only be a more refined set of probabilities. Down syndrome can be reliably predicted by amniocentesis, Samerski says, but even then one still has to decide the meaning of the prediction in the dark, as it were.

SS: Down syndrome is one of the cases where the information is pretty certain — whatever that means. That's another question. The geneticists can tell you your child has trisomy 21, so when it's born, it will certainly get the diagnosis Down syndrome, but children with Down syndrome can be very different. It's the prediction of a diagnosis, but it doesn't say much about the child that you really hold in your arms later, I would claim. There are a few other predictions that are certain, but in many cases, it's all about risk and probability. For example, for a long time, cystic fibrosis was considered one of the exemplars of a monogenetic disease that you could predict.

DC: Monogenetic: located on one gene?

SS: Yes, located on one gene. The idea was that there's one gene, and that's the cause for the disease. But after years of more research, they have found that it's not so simple. A doctor here told me he sees two siblings who have the two so-called severe mutations on the gene that was considered the cause of this illness. He said they're fine. They're now twenty-something. They do sports. They're fine. Even in that case, from the genetic tests, you can't tell with certainty if a child will be severely sick. You have to give parents risks, probabilities, which say nothing about the concrete case.

DC: The difference between the concrete case and the abstract risk is what Samerski's research is all about. She thinks that they belong to entirely different orders. An abyss separates the world of experience from the world of statistical probabilities. Genetic counselling and genetic screening merge these orders, and the result, she fears, is a denaturing of experience.

SS: The woman is asked to decide about a child that is not yet there. She can't look into its face, she can't hold it in her arms, but she gets some averages, like the IQ. The decisions she's supposed to make are based on the predicted IQ, on the average outcome of children with Down syndrome, and so on and so on, but she doesn't know her concrete child. Then she's asked to decide if she wants to keep the pregnancy or not, and ninety-five per cent of those ones who get that result abort. You could say, of course, it has an effect that is somehow also eugenic, because there are not so many children with Down syndrome born anymore, but I don't think that those women necessarily have some eugenic conception of their family or of the general population. I think it's very difficult to keep a pregnancy and to be pregnant with a child that is already diagnosed, that is already put into all kinds of statistical populations. The child's future is already somehow not open anymore, and I think that's one of the difficulties. I assume that many women would maybe be very happy with the child when they would deliver it, but it's somehow very difficult to still be pregnant, to be in good hope, with a child that is completely checked through and diagnosed and considered abnormal.

DC: Genetic screening imposes a decision; it turns the mother-to-be into the manager of a pregnancy. For some of the people Samerski observed, there was no difficulty in this. It was merely an extension of an already familiar and comfortable stance. For others, it was shattering.

SS: I once attended two cases where amniocentesis was performed. In the first case, the woman came with her husband, and when the physician asked her what she would do if the result did not give the green light, she said, "I don't want a handicapped child. I will abort. That's clear." From the way she entered that room and everything about her, you could see that she had incorporated that attitude. Check the child. If it's not okay, it's going to be aborted. The next woman was already in tears by the time she entered the room, and she said, "Why is all this happening? I was feeling so well, and I thought, okay, let's just do this triple test, then I'm through with it. I'm feeling good, and I think the child is healthy, so the test will just confirm this feeling. Now it didn't." She was constantly in tears, and you could see she didn't really want to be in the position she was in. It was the opposite case — a woman who wanted to be in good hope and had been in good hope and was just somehow forced into that system. Then she said to herself, okay, as a modern woman who wants to be responsible for her coming child, you have to do certain things, undergo some tests to be responsible. Then she was caught by that system. These are two extreme cases, but I think that for most women it's when they finally have to make the decision about whether to keep the pregnancy or not that it all becomes real. Before that, it's all about numbers and fantasies, but when you have to decide whether to keep the pregnancy or maybe abort, then it becomes real.

DC: Women undergoing genetic counselling are invited to make a decision about their pregnancy. The counsellors are generally quite scrupulous about not giving any advice. We will give you information, they say, and then you must decide. In this sense, Samerski thinks, genetic counselling enacts what she has called "the paradox of taught self-determination." The decision may be impossible, but making it shows that you are free.

SS: In general, in our society, the idea of making autonomous decisions is considered the epitome of freedom, so you are free if you can go to the supermarket and choose between different shampoos and go to the travel agent and choose between different countries to go to. You're also considered an informed client if you can go to the doctor and choose between different options and tests. I always get in trouble with my physicians when they ask me, what do you want?, and I say, I don't want anything here. If you consider something necessary, please tell me. But I don't want anything. Then they tell me, maybe if this calms you down, we can do it. But I don't want anything to calm me down. I want something that makes sense. That's a contemporary way of dealing with clients. People are urged to make decisions, and there are more and more counselling services that tell people to make their own decisions and tell people how to make these decisions and give them the right information to make decisions and so on. I think that one of the symbolic effects of this taught decision-making is that it asks people to feel responsible for what happens to them. In a world where people have less and less influence on their surroundings and their society, they are asked to feel responsible individually for their own fate, for what happens to them.

DC: "Genetic counselling," Silya Samerski has written, "is . . . an educational ritual which prepares citizens for a new kind of responsibility." The client in genetic counselling is responsible for things that are imputed to her by expert professionals — for a certain constellation of genes that generates a risk profile — and not for anything that is actually within her personal grasp. This is the paradox of the client's situation.

SS: Those women are asked to feel responsible for something they can't be responsible for, and I think that's why genetic counselling is so much a paradigm for a more general development or tendency. You can't be responsible for the outcome of your pregnancy, but all these tests and decisions that these women are urged to make — they make you feel responsible. You can't influence the health of your child or the makeup of your child. You can make these horrible decisions of aborting or keeping the pregnancy, despite all the risks that are attached to it, but you can't be responsible for the outcome of your pregnancy.

DC: Genetic counselling puts women in an impossible position: they feel responsible for what they cannot, in fact, control. What makes this impossible situation somehow plausible is the idea people are encouraged to form of genes. The popularly understood gene gives a body to what would otherwise be mere statistics. It incarnates risk, and that, Samerski says finally, is its ultimate significance.

SS: If there was not this idea of the gene, it would be much clearer that all these numbers and all these risk figures don't have anything to do with this woman. The gene is somehow the bridge between these abstract statistical figures and the concrete person. It merges the statistics and you, or the statistics and the concrete person. It bridges the gap that is there. So if a risk is said to be a genetic risk, it looks like there is something already pre-programmed in your body. This disease is somehow already latently there even before it breaks out. In fact, risk only tells you that you have been put into a certain population. Risk refers to the frequency with which something occurs in this population. But it gives you the impression that it's genetic, so it's somehow already inside of you. It's like a pre-programmed fate, and I think that in this respect the gene has a very important symbolic function. This is true in public debates as well. The gene delivers the kind of person and the kind of body that fits into a risk society.

The Price of Metaphor

RICHARD LEWONTIN

If I take a sheet of paper that you give me with some words on it, and I take it upstairs to the Xerox machine, and I put it in the machine and I press the button, and out comes a copy, I don't say that piece of paper copied itself. I don't say it replicated itself. I say a copy of it was made by the copying machine. That's exactly what goes on in the cell in making DNA, new copies of DNA. DNA is no more powerful than that piece of paper you gave me to make a copy.

— Richard Lewontin

In 1990, American evolutionary biologist Richard Lewontin delivered CBC Radio's annual Massey Lectures under the title *Biology as Ideology: The Doctrine of DNA*. In his lectures, Lewontin argued that science had replaced religion as what he called the "chief legitimating force in modern society." Science sanctions the existing social order, he claimed, by telling stories about a universal "struggle for existence" or about how we are all blindly programmed by our selfish genes. These stories, in Lewontin's view, constitute the ideology of biology and demand careful and continuous scrutiny.

Many years ago, two pioneers of the new science of cybernetics, Norbert Wiener and Arturo Rosenblueth, wrote a sentence that Richard Lewontin has often quoted: "The price of metaphor is eternal vigilance." It sums up a lot of what Lewontin has had to say over the years about the misuses of science. Lewontin is a distinguished geneticist and population biologist who has added to his science in fundamental ways, but he has also been a philosopher of biology, and there his contribution has been to carefully delineate the limits of his science. He has criticized the extension of scientific ideas into social domains where they don't apply, and he has kept a wary eye on the metaphors which carry scientific findings into everyday speech and everyday understanding. I interviewed Richard Lewontin in 2007 at Harvard's Museum of Natural History, where he's now professor emeritus and where for many years he had his lab. He spoke about the danger of transplanting ideas developed in one field into others where they don't fit.

RICHARD LEWONTIN

Every set of phenomena has its own domain of operation. I'm strongly opposed to carrying over from one domain of phenomena to another a model which works in one domain but not in the other. I wouldn't want to carry over nuclear physics to the organization of society, nor even to biology. I don't want to carry over Darwinian evolutionary principles to historical changes in human societies. They're not the same. My generalized view of the world is, yeah, it's easy to take a well-established theory and try to slap it on some domain it wasn't designed for, but that's not the way to find out the truth. That's why I'm so disdainful of people who talk about cultural evolution or linguistic evolution. Why evolution? Culture has a history. Language has a history. But the minute you say it's evolution, there's a certain implication that the structure of Darwinian evolutionary theory can be laid onto it. As an epistemologist, I dislike that.

DAVID CAYLEY

Lewontin is opposed to a wholesale transfer of terms, models, and metaphors from one field into another, but his ban is not absolute. He recognizes that ideas from one domain can sometimes play a suggestive and creative role in another. He gives, as an example, the work of Gregor Mendel, the Silesian monk from what is today the Czech Republic, who, in the 1860s, discovered the basic mechanism of heredity by experimenting with pea plants.

RL: It has had constructive effects, even though you have to be wary of it. To take an example that isn't often talked about, except in the new understanding of Mendel by some Czech historians of science, the important thing about Mendel was that he was a physics student. He was recruited into the monastery at a time when physics was becoming particulate and molecular, and that was the way he saw the physical world. I think, though you can't prove it, that that had a very powerful influence on his development of a particulate theory of genetics, which is what Mendelian theory was. He wasn't working with blending inheritance. He was working with yellow versus green, wrinkled versus non-wrinkled. He developed this idea of factors, as he called them, which separated in the formation of sperm and pollen and eggs and came together again after fertilization but didn't blend with each other. They maintained their individual identity, and they separated again when that organism formed pollen or eggs. The whole dance of evolution, the coming together of these factors that are being mixed within the individual but not themselves mixing with each other — maintaining their individual identity — I think he got from his notion of particulate physics. So I don't want to claim that one never can make progress by carrying ideas over from one domain to another — that's what metaphor means, it means to carry over — but you have to remember what Wiener said: "The price of metaphor is eternal vigilance."

DC: Vigilance needs to be exercised, Lewontin believes, because it's easy to forget that metaphors are metaphors, that they are provisional and limited comparisons, not literal descriptions. To avoid confusion, one has to know which aspects of the metaphor fit and which don't.

RL: I learned in physics, when I was a student, what's called the billiard ball model of molecules — the idea that they're like billiard balls, that they collide and bounce off each other. That model is fine because we don't think that molecules come in yellow, blue, green, and red. We don't think that they have little numbers on them. We don't think they go click when they hit. We abstract from that model a particular feature of billiard balls, namely that they have elastic collisions. That's fine, because we already know that much of the metaphor is wrong and irrelevant, and we're not foolish enough to import it into the thing we're trying to describe. The trouble with most metaphors in biology is that the people using them don't know what to import and what to leave out. You already have to know a lot about the thing you're metaphorizing before you know what's relevant about the metaphor and what isn't. You can always say, well, genes are like a code, provided that you really know what's going on with genes and you don't think that they're entirely like a code. For example, we say DNA makes copies of itself. It's self-replicating, but that's wrong. I mean, it's completely wrong. DNA does not self-replicate. DNA is . . . but now I'm going to use another metaphor. I have to. I can't talk without metaphors. That's very important to understand. You can't get rid of metaphors. You just have to be conscious of them all the time. So DNA is manufactured by a cellular machinery made up of a lot of different enzymes. Little bits and pieces are put together. The enzymes act to link the DNA up. The DNA is a copy of a previous DNA molecule in the cell. But that DNA doesn't copy itself. It is copied by a machine, which sometimes makes mistakes, by the way. To carry the point about the metaphor to its end, if I take a sheet of paper that you give me with some words on it, and I take it upstairs to the Xerox machine, and I put it in the machine, and I press the button and out comes a copy, I don't say that that piece of paper copied itself. I don't say it replicated itself. I say a copy of it was made by the copying machine. That's exactly what goes on in the cell in making DNA, new copies of DNA. DNA has no more power than that piece of paper you gave me to make a copy.

Now, DNA itself, as part of its code, tells the cell machinery how to make further copies of the cell machinery. In other words, the machinery of the cell makes DNA, but DNA has — in its code, if you like — a recipe for making the enzymes that make DNA. So DNA is both the cause and the effect of the perpetuation of that machinery. It has a causal role because it contains the specifications for some aspects of that machinery, but the machinery is necessary in order to make new DNA.

The other thing said about DNA is that DNA makes proteins, right? But it doesn't make proteins. That's wrong. The chemical sequence of DNA is the recipe that says which amino acid should be hooked up to which amino acid in making a protein. Yes, it is, that's the recipe, but a recipe doesn't make a cake. A recipe tells the maker what to put in. So DNA doesn't make the proteins. The cell reads it off, so to speak. It says, well, I should take this amino acid and stick it to that one, and I should take the next one and stick it to that one. I should take the next one . . . so it's the cell machinery that is making the sequence of amino acids. But even after the cell machinery has made that sequence of amino acids, that sequence is not a protein. A protein is a modified, folded, somewhat cut up, somewhat attached set of amino acids that have long since left that recipe behind. When the sequence of amino acids is made in a long chain, that sequence folds up into a three-dimensional structure which is the protein, and the folding of that structure is not specified a hundred percent by the sequence of amino acids. People often say, if I know the amino acids, then the folding is given. It's not true. That's one of the agonies of modern molecular biology and molecular physics and chemistry. If I tell you the sequence of amino acids in a protein, you cannot predict the folding of that protein. You may be able, with sufficiently sophisticated computer programs, to predict that there would be one of the following twelve stable foldings, but you can't tell which of those twelve stable foldings will in fact occur because the folding process is itself a historical act. The protein folds up bit by bit — preliminary folding and then further folding — in a certain cellular milieu, at a certain pH, with certain other molecules around. Which specific folding occurs, out of the twelve possible foldings, is not coded in the amino acids. It's a result of the environment in which the folding has occurred.

A famous practical example of this is insulin. Insulin doesn't come from pigs anymore. Insulin is now made in vats from the human insulin gene. It was originally put into bacteria, but when they did that, they got the amino acid sequence all right, but they didn't get insulin. It misfolded, and it had no physiological activity at all, even though it had the right amino acid sequence. The scientists at Eli Lilly screwed around until they found just the right way to cut it up and repackage it and fold it so it made insulin, even though the amino acid sequence looked the same. That's a classic example of why folding is so important. There's a whole branch of structural studies which is now devoted, and has been for some years, to "the folding problem," as it's called.

Biology, in my view, is becoming more and more dominated by this issue of how proteins fold up and how proteins find each other in the cell. You know, there are only three copies of that protein, and two of that, and seven of that. They've got to be in the right place at the right time, and they have to rotate around to match each other. We have no understanding of how that happens. That's the real biology at the cellular level. Cells are not like flasks full of a chemical. We don't have 10^{23} copies of things. We only have three or four of one and five or six of the other. That means there's a lot of random noise in cells — random molecular noise — and when the cell divides, you won't have exactly equal numbers of those molecules in the two new cells. One cell will have five, the other will have three. Then that cell has to wait around awhile until it can make more copies in order for it to do its job. For example, if I take a single bacterium and I put it in a big flask full of nutrient, after about an hour the bacterium will divide into two bacteria. But those two bacteria don't divide simultaneously an hour later. First one divides, then the other one divides, and if I wait another hour, one of those will divide, and then later another, and then another. Pretty soon, every cell is dividing at a different instant. Now, why is that? They have the same genes. After a few divisions, there haven't been any mutations. They're living in this mixed soup where their environment is identical. What's happened is the effect of developmental noise. When that first cell divided into two, the contents of those two cells were not identical. They both had the same DNA, but one had five copies of a certain molecule and another one had three copies of the same molecule. That's not enough for cell division and it had to wait around. That's a very important phenomenon in biology.

If you look at your fingerprints, your fingerprints are not the same on your right hand as they are on your left hand. Why is that? You got the same genes on your right and left side. Both sides had the same environment when you were a little baby, folded up in your mother's womb, and the fingerprints were being formed. You can't talk about the environment of the left hand and the environment of the right hand. Yet the fingerprints are quite different. And you can reproduce what happens with your fingerprints. Take a sheet of paper out of a new batch of paper you've bought at the paper store, put it between two pieces of wood, and slowly push the pieces of wood together until they crumple the paper up. Now take a second piece of paper from that same package, put it in there, and again push the two blocks of wood together until that piece of paper crumples up. Now look at the two pieces of paper. They won't be identical in their crumpling, absolutely not, because there are minute differences in the thickness of the sheets and in the way the strands of wood lie in the paper. Those little differences make all the difference in how each sheet folds up. That's the story of development of cells and tissues. People don't study that because it's very inconvenient, but it's true.

DC: I want to go a little farther with the question of the misuse of metaphor . . .

RL: Okay, development. How about development? There's a metaphor.

DC: It's true.

RL: Development means literally to come out of an envelope. Look, when you take a picture, or used to take a picture in the old days before electronic pictures, you put it in a bunch of chemicals called a developing bath. Why is it called a developing bath? Because the picture's already immanent in the film, and it develops. But what develops is just what was there already. It unfolds. It unrolls. You have to look at it in other languages. In Spanish the word to develop is *desarrollo*, meaning an unrolling. In German it's *entwicklung*, an unrolling of a ball of twine. We always have this idea that development is something that's already immanent in the egg, from which it just unrolls and unfolds. That's what the DNA-ologists are trying to push on us: what you are is already immanent in your DNA, in the fertilized egg, and you're nothing but the unrolling and unfolding of that genetic program. But that's not true. We know it's not true. Even identical twins are not identical. They don't have identical fingerprints, for example. They have a lot of differences.

A famous Canadian case which I would like to talk about for a moment is the Dionne quintuplets. They came from rural Ontario. Their father was not very well off — he already had a large family. I have a picture of the Dionne quintuplets, standing one next to the other, at the age of three or four, all lined up, with identical hairdos, identical dresses, and identical shoes. You, as a Canadian, may remember that they were on show in a kind of zoo. Dr. Dafoe, who delivered them, convinced the province of Ontario to invest money in a public display of these children playing with their toys, identically dressed, and so on. They were brought up identically, and they had identical genes, yet they didn't turn out identically. Two of them had a religious vocation; one of them failed at that vocation. Two are dead. Three are still alive. Two got married, or three got married. Two had children. If you look at them now, you can see they're sisters, but they're not identical. They had different life histories. Their life histories were as different as the lives of any five girls brought up in a rural, not-very-well-off Ontario family, even though they had identical genes and an identical environment up until the age of sixteen or so. That's developmental noise.

DC: Can we talk for a moment about the ideological uses of biological science?

RL: Whatever you like.

DC: Why is this deterministic, you-are-your-genes view preferred?

RL: Well, biologists prefer it because it gives them a simple view of a world in which they have all the answers. I once heard a very famous molecular biologist say, if you give me the DNA sequence of an organism and a big enough computer, I can compute the organism. Biologists love that because that validates their whole operation. All I have to do is sequence the DNA in people, learn the rules of translation of DNA into proteins, and learn how proteins get together, and then I have the complete machine. I've solved all the problems of biology. Indeed, I ought to be able to make an organism. Biologists are ambitious to be like physicists. After all, if I ask you who is the greatest scientist who ever lived, you're likely to say Einstein . . .

DC: . . . if I don't say Isaac Newton.

RL: If you don't say Isaac Newton. Fine. I'd prefer you to say Isaac Newton, for my purposes, because the reason you say Isaac Newton is that he had a set of laws which were supposed to apply everywhere to everything. The model you and I learned in school for science is universality. The scientist is more important the more universal the thing that scientist has discovered is, and less important the more specific to this one particular instance his discovery may be. Every scientist is raised with the image of being, if not a Newton, at least Newtonesque, of doing something which is general. Molecular biologists are just like everybody else. If you want to be famous, you do what Watson and Crick did. You find the universal code for how DNA is made — of course, it's not quite universal, but close enough.

If you just say, well, I'm going to spend my life studying the life history of this frog, you don't become very famous and don't make a lot of money. It's a feedback from that training and that ideology. If that's the way a person is going to become famous, and those are the most important people, then we should put the most money and time into people who are doing that, and we must believe that it's possible. If I say to you, I think the human species will become extinct before we have a reasonable understanding of the central nervous system and the brain, you will regard that as terrible. I think it's true, by the way, and it doesn't bother me. That is to say, I think this is a material world. Everything in it is material. Everything can be given as an interaction of material causes at some level, including the quantum level. But that's not the same as saying we will know the truth about all things. The answer is no. The human species will become extinct before we know the truth about lots of things. You have to distinguish between knowing the truth and there being a truth. I think the material nature of the world is a truth, but we don't know it, and we can't know it. It's too much to know.

DC: This idea that physics is the model, and a certain idea of physics as well . . . was that aggravated in biology by the migration of physicists into the field?

RL: I think that your word "aggravated" is the correct one. It was not introduced into biology by those people, but certainly the present model of molecular biology comes from the fact that the present generation of molecular biologists are the inheritors of a movement into biology by physicists and by a certain kind of chemist, but especially physicists, at the end of World War II.

Max Delbrück was a physicist, and many of the people who came into biology around Max, and around other people, came in through physics, absolutely. They've done great work, but they've presented a model, and the early ones were abysmally ignorant of biology. I remember going to Cold Spring Harbor Symposium and hearing one of those former physical chemists or physicists talk about one of the mechanisms of chromosomes, and he was just going on about nonsense. The reason it was nonsense is because he didn't know the fundamental phenomenology that he was talking about. Now that's past. I mean, people in molecular biology now know what they have to know. So, yes, you're quite right, that model of physics has had an exaggerated importance, but it was not introduced by physicists coming into biology. On the contrary, physicists came into biology because they ran out of steam in physics, and they looked for new worlds to conquer. They didn't dream for an instant that it would be any more difficult than physics. They would still get Newtonian and Einsteinian and Diracian — and so on — laws in biology.

DC: Did it cause a split in the field?

RL: It did cause a split in the field. It caused a split, in fact, in the institution in which I now work. When I was a student, and when you were a student, we had a biology department. That biology department consisted of evolutionary biologists, who were interested in classifying organisms and talking about evolution, and what I call functional biologists, people who cared about how cells worked, how development worked, and so on. Then along came molecular biology. It couldn't be quite fitted into the biology department, and so they formed a separate group of people doing biochemistry and molecular biology. We had three different groups. Then the biology department split into two parts — the organismic and evolutionary biology department and the cellular and developmental biology department — because suddenly developmental biologists didn't want to be in the same department with all those butterfly collectors. Those two groups separated, and then there was the third group, the molecular biologists. Now, we once again have them coming together in a certain way, but not entirely. We still have separate departments, and we now have three departments.

DC: This idea that they didn't want to be in the same department as the butterfly collectors, it sounds offhand, but perhaps . . .

RL: No, it's true. They said that.

DC: They were doing science . . .

RL: . . . they were doing real science . . .

DC: . . . not natural history.

RL: That's correct. Natural history is not science.

DC: It's just collecting.

RL: It's just collecting. I was recruited to Harvard as a weapon in that struggle, because I came to Harvard as an evolutionist who studied molecules. I was the first one here, and they gave me a whole floor, a new laboratory wing of the museum — I was in the museum, I'm a professor of the museum — and asked if I would please train all the classifiers of plants and animals in these new techniques so that we would be able to hold up our heads in biology.

DC: And?

RL: And it worked. Evolutionary biology is now recognized as having a proper molecular branch and is no longer looked down on — although I have to say that there may still be some looking down on the part of evolutionary molecular biologists on the people who do nothing but classify organisms.

DC: Well, we're surrounded here, where we're sitting in the museum, by partially reassembled dinosaur bones.

RL: And they look down on people who collect dinosaur bones. Yes, they do. Because they have come to believe that the ultimate truth about biology is molecular, even though they're evolutionists. There's still a split, but it's along a new suture line. It's within evolutionary biology, but it's not as bad as the old one, because people who classify organisms do it using molecular characters. They have laboratories. They sequence DNA. They no longer classify organisms by how many hairs they have. They classify them by their DNA. There has been a reunion, and now I would say that the split between those who are interested in what kinds of organisms there are and how they're related to each other and people who do molecular biology is minimized. It's not serious anymore.

DC: Richard Lewontin was recruited by Harvard as a weapon in a struggle that was going on in the biology department. This was in the 1970s. One of his jobs was to train the old guard in the department, the classifiers of plants and animals, in the new techniques of molecular biology. His qualification was the breakthrough work he had done at the University of Chicago in the decade before his appointment at Harvard. Lewontin had done graduate work in the 1950s with Theodosius Dobzhansky. Dobzhansky, a Russian émigré, was a major contributor to what Julian Huxley in the early 1940s called "the modern synthesis" in evolutionary theory. This synthesis put Darwinism together with the new science of genetics, showing how natural selection worked on genetic variation within populations. But, at the time Lewontin was studying with Dobzhansky, there was still no way to find out in detail just how much genetic variation there actually is. That was the problem Richard Lewontin set himself.

RL: In Dobzhansky's day, nobody knew how variable single genes were in nature. Nobody knew, and the first part — or the middle part, really — of my scientific life was devoted to answering that question. How much genetic variation is there, gene by gene, in any old species you care to name? We found a method to do that using electrophoresis of proteins, moving proteins in gels. I met a guy in Chicago, a guy named Jack Hubby, who knew how to do that but didn't know what to do with it . . .

DC: . . . moving proteins in gels?

RL: Yeah, you grind up an organism and you extract its proteins, and then you squirt them onto a slab of Jell-O, and you turn on the electric current. You put a positive pole at one end of the Jell-O and a negative pole at the other, and you turn on the current. All of the proteins migrate through the Jell-O, but they migrate at a rate that depends on their composition, on their amino acid composition. If there's a slight difference between the proteins made by the two copies of a gene you have, you'll have two different bands on this Jell-O. Jack Hubby was using that to demonstrate that one species looks different from another species. I met him, and he told me about his work, and I thought, my God, there's the method I've been looking for my whole life — my whole life, ten years or whatever it was. So I moved to Chicago, Jack and I set up shop together, and within two years we had the answer. It was so easy to do — any damn fool could learn the technique in a couple of weeks — that it suddenly converted the whole field to what I called "find 'em and grind 'em." You took every conceivable organism you could find — I mean, thousands of different species, from bacteria to fruit flies to plants to people to animals of all kinds — extracted their proteins, ground them up, put them on this Jell-O, and turned on the current. Then, by specific staining of each protein, you knew which gene you were dealing with. You could distinguish the product of one gene from the product of some other gene. We found out a tremendous amount about how much genetic variation there is in natural populations of all kinds of organisms.

DC: Presumably you wouldn't have needed to grind up whole people . . .

RL: No, a fingernail or a little blood will do. That's right. You don't have to grind up a whole mouse, but you do grind up a whole fruit fly. You don't have to grind up a whole plant — just take off a leaf.

DC: What did you understand, finally, from that work?

RL: In the typical sexually reproducing organism — whether it's a plant or an animal or humans or anything — about one third of its entire complement of genes are what we call polymorphic. You have more than one variety of that gene floating around in the population at some reasonable frequency. By the way, nobody does this with proteins anymore. It's done with DNA, and DNA gives the same answer.

DC: What are the implications?

RL: Well, you can't evolve unless there's genetic variation in a population, so the amount of variation tells you we have a tremendous potential for evolution. There's plenty of genetic variation out there. If natural selection pressures changed, there would presumably be new forms that could be selected. But all it tells you is the potential for evolution. Period. It doesn't tell you how much evolution is now going on.

For example, I don't know anything about where Dobzhansky's species of drosophila lays its eggs in nature. I don't know how much struggle there is between larvae for survivorship. I don't know how a female finds the right place to lay an egg. I don't know how a male finds a female. All of those things matter in determining how many offspring you have. You've got to find a mate, you've got to find a place to lay an egg, the eggs have to hatch, the larvae in mature stages have to survive. The flies have to hatch out. Nobody knows anything about that. Nobody knows it for mice. Nobody knows it for fruit flies. Nobody knows it for frogs. That's the problem — to find model organisms in nature.

DC: Biology still knows very little about the life histories outside the lab of the creatures it has used as its models. Even so, Lewontin's work was a major advance in his field, and it had an interesting spin-off. He was able to show, in a paper he wrote in the early 1970s, that there is much more genetic variation within human groups than there is between groups.

RL: I had to write a paper for a volume on evolutionary biology, and I thought, what am I going to write a paper about? I went on a bus trip somewhere, and I took some books of data with me and a little hand calculator and a table of logarithms. I thought, oh, here's something you can do when you're on a bus, that you don't have to be in the lab to do. I spent the whole bus trip reading off the data and calculating the results, and it was quite amazing. I found that eighty-five per cent of the variation was between individuals within groups. Other people then followed suit with independent samples and with new genes that I hadn't looked at and so on. My real amazement, being an experimentalist, was that they came out with almost the same answers. I mean, we don't have results like that in biology. I said 85.4 percent was within groups. The next person found 86.2 percent. The next person found 84.7 percent. And the last study done — it was done on DNA, which we didn't do — found that 85 percent of the variation is within groups. So I can't think of a number that's more well established in biology than that. Eighty-five percent of all human genetic variation is certainly within local populations, and I mean really local populations. Within a local linguistic group. Within the French. Within the Italians. Within Kikuyu — not just within Africa, but within Kikuyu.

DC: Richard Lewontin's finding and its subsequent confirmations demonstrated a remarkable underlying consistency in the human species. This had large political implications, insofar as it showed that race is a category with no deep biological foundation. But Lewontin had not set out to make a political splash. In fact, he says, when he undertook his study he expected the opposite result.

RL: I really began on the assumption that when I looked at the all the data it would turn out that most of the variation in human groups was between groups and not within groups, and I was surprised to find that it's mostly within groups . . .

DC: Well, it is counterintuitive, because superficially, at least, we look rather different.

RL: So the question is, why are most of the differences that do exist between groups superficial differences? Is that an accident? How come it's skin colour and hair form and nose shape and height and that stuff which are the clear differentiators of geographical populations? Why are the people in Africa one skin colour and the people in Europe another, whereas for most our genes you can't tell them apart?

DC: You don't really think I'm going to tell you, do you?

RL: Well, I don't know the answer. Nobody knows the answer. But it's a fact. There is one suggestion for that answer, and that is that most of the differences between geographical groups are a consequence of what's called sexual selection. Darwin talked about that. The argument is that it's a consequence of the fact that some people look the way you want your mate to look, and so those people are preferred as mates. Now, why would some people in Africa prefer dark-looking people, and some people in Europe prefer light-looking people? One could only say — and now you're making it up, you see, 'cause you don't have any evidence — because just by accident, when those populations were very small, the big cheese in the local group happened to be darker than average, and all the women wanted to have a husband who looked like that. Or the most sexy woman looked a bit darker than the others. So people would choose as a model for their mate something to do with power or how good that person was at doing something else. Just by accident, that would be different in Europe than it would be Africa. It had nothing to do with whether it was Africa or Europe. It had to do with there being different populations not in communication with each other. Why do people in Asia have the eye shape they have? Why do they have the skin colour they have, which is different from both African and European skin colour? There are models of what you want your mate to be like; that's the theory of sexual selection. Remember, I just made up those stories. I have no knowledge at all of why people in Asia look the way they do and why people in Africa look the way they do. None. And nobody else has, either. They just do.

DC: Richard Lewontin's work on genetic diversity and its distribution was certainly his signal contribution to his field, but he has also had a lot to say about the proper interpretation of the theory of evolution. For example, in 1985 he published a collection of essays written with his friend Richard Levins called *The Dialectical Biologist*. The word "dialectical," for Lewontin, refers to a way of understanding causes and effects as interpenetrating, an approach he recommends to his fellow evolutionists. He gave an example earlier when he described DNA as both a cause and an effect of its cellular environment. However, his colleagues, generally speaking, do not think in this way.

RL: Evolutionists, by and large, describe evolution, roughly speaking, as the adaptation of organisms to their environments. So you have the environment — there it is — and the organism fits into it. The very word "fit" has that metaphorical notion of a hole into which you can snuggle and . . . fit. But that's wrong. Organisms do not fit into pre-existing ecological niches, as they're called. Organisms change. They change the world around them. They create their environments. They construct their environments out of the bits and pieces of the world in ways that are particularly — if I may be cruel about it — dictated by their DNA. In the process of choosing bits and pieces of the world to interact with, they change those bits and pieces. Plants put down roots, and, when they put down roots, they put chemicals into the soil which affect how the next plant will grow. They put both nourishing things and poisons into the soil. They break up the soil. They make it physiologically different.

Human beings, and many other living organisms, are surrounded by a layer of warm, moist air which is moving up over our bodies and off the top of our heads. You can see this with the right kind of photography. We are living inside that warm, moist shell. We're not living out there in the atmosphere, and that's why we have the so-called wind chill effect. When the wind blows, it blows away this layer of warm moisture and suddenly we're really out there, really in the atmosphere. Every organism at every moment is manufacturing aspects of its environment, altering aspects of its environment, which affects both other organisms and itself. As the organism evolves, its environment evolves along with it.

This is not to say that there are no external forces affecting evolution. The sun sets and rises without reference to any organism. Spring follows winter without reference to any organism. Cyclones, weather patterns . . . well, they're not quite independent of any organism because it appears that we are doing things in the world which are changing those patterns, but on a small scale, whether it rains or not doesn't depend on what I do today. So there are external forces, also, but the way they impact on the organism depends on the activity of the organism. Thus organism and environment interpenetrate. They are both the causes and the effects of their mutual evolution.

DC: The theory of evolution has been very hard for people to swallow, evidently.

RL: Well, do you mean the theory of evolution or the fact of evolution?

DC: Well, okay, the fact of evolution.

RL: The fact of evolution has been very hard for some people to swallow. Once they swallow the fact of evolution, they don't have any problem.

DC: But the theory can be expressed in various ways. The way you've expressed it is rather different than the way it's sometimes expressed.

RL: Or the way others of my colleagues express it.

DC: I'm just saying that it might be more congenial in the form in which you've expressed it than in its more mechanical formulation . . .

RL: But I also express it in a perfectly mechanical way. The interactions between organisms and their environments involve some kind of physical causation, and I can study them by physical means . . .

DC: Yes, I understand. I'm not saying it's a non-materialistic theory, but it is a holistic one, in a certain sense.

RL: Well, but it's not holistic in the obscurantist sense that the world is one big unanalyzable whole. I reject that completely. The world is analyzable. It's analyzable into bits and pieces, provided you remember that your process of analysis determines, in part, what the bits and pieces are, and, in part, the bits and pieces are interacting with each other constantly.

DC: Holistic, as I learned, is not a word Lewontin favours, but it's certainly fair to say that he has been an opponent of reductionism in biology — reductionism being, essentially, the idea that everything can be reduced to some fundamental law, level, or structure. He has been especially vigorous in opposing what he has dubbed the Doctrine of DNA, the view that genes determine everything. This doctrine has now largely exhausted itself in academic biology, he says. Attention has switched from the genome to what is being called the proteome — to the complex of protein molecules that are made from instructions found in DNA and that then enter into the many processes that go on in our cells.

RL: DNA doesn't have enough information in it. We have to understand what the proteins look like, and we have to understand their distribution in cells. That's one point. The second point is that most people in the world are not dying of defects in their DNA. They're dying of under-eating and overwork — or in some places, perhaps, overeating and underwork — but most of the world's people are dying of under-eating and overwork. If you're really concerned with human welfare, you won't look at their DNA. I think that seems quite clear. If you want to understand biology, you have to move beyond the DNA sequence to the proteome, as it's now called — all the proteins, and how they're folded, and where they are, and how many molecules there are, and how the sub-cellular structures are formed, and what passes through the pores in the cell membrane. There's a lot of biology like that being done now. Biology is moving into the ultrastructure realm, the realm of cell structure and molecular structure, and leaving behind the DNA business.

DC: "Science is a social institution." With this sentence, Richard Lewontin began his 1990 CBC Massey Lectures. He went on to try and demonstrate the integration of science with the existing social structure. That's been his view throughout his career, and it's a view that he thinks has won wide acceptance during that time. He points, for example, to the way the history of science is now generally told.

RL: The way science history has been written has changed in my adult professional lifetime. When I was in college and became an adult, the history of science meant biography. It was the great-man theory, or what we sometimes call the Cleopatra's Nose Theory of History: if Cleopatra's nose had been misshapen, would Roman history have been changed? Everybody did great-man history. There were hundreds of books about Darwin. People still write about Darwin, but the reason for doing so and the underlying structure of these books has changed in the last thirty or forty years. The doing of science is now seen as a social phenomenon with a communal setting of agendas. People have begun to recognize, although I feel not enough has been done with it yet, the importance of the notion of the peer group. Science, as it's done now, is very expensive. Where does the money come from? It comes from governments. How is the money awarded to scientists by the governments? They have committees. Who are these committees? These are committees of your scientific peers. You don't put on the committees people who are not recognized by the scientific community as peers. But that has a very powerful effect on what science is done. Why are they recognized as your peers? Because they do the science that people think is important. It's very hard to do things that are not part of the agreement of one's peers on what's important. In that way, the existing consensus tends to replicate itself. There's too much peer review, in my view, and not enough chance for people to do things that are not recognized. I was on those committees for years. I mean, I know how they work. Peers give out professorships. How is it decided whether you're going to get a job? You get a job by the evaluation of your peers and by what the people in the department think your peers think of you. How do you get prizes? Science is organized as a self-reinforcing peerage. Once in a while, people break into the peerage with something new — it's not totally static — but the people who get to do the most famous things do the things that are already on the agenda, even if they haven't been done. Watson and Crick did what everybody thought had to be done, namely that we had to know what the structure of DNA was. They didn't invent the idea that the structure of DNA was important. Not only that, but they used the technique which had been used for years and years and years by people who studied the structure of molecules — X-ray crystallography. They brought a historically well-used technique to bear on a problem that everybody said was *the* problem. Why did they do it? 'Cause they were very ambitious guys, and if you're very ambitious, you do the thing that everybody says ought to be done.

DC: Would you say the same of Darwin?

RL: Yes. Darwin was already a fellow of the Royal Society when he started to work on evolution. He was well known as a geologist. He was already part of the scientific establishment when he came back from his trip around the world, and he was urged by other members of the Royal to hurry up and get this stuff into print because there was this guy Wallace who was putting it together, and, if he really wanted to beat out Wallace, he better get that book going. Darwin was already a big wheel in science.

DC: Darwin worked on the problem assigned to him by his time and by his scientific milieu, and the same can be said of Darwin's contemporary, Gregor Mendel, whose work became the foundation of genetics. Mendel, the obscure Silesian monk, puttering with pea plants in his garden, is among the hoariest of scientific myths, but what the new Czech historians of science are finding is that Mendel was also working in a compelling social context.

RL: Mendel didn't start out as a monk. He was a student of physics, and his professor sent him to the monastery. The monastery at that time was doing a lot of scientific work. In fact, the bishop threatened to close the monastery up, because they spent too much time doing science and not enough time praying. They did work on the weather, and they did a lot of work to help the fruit orchardists and the farmers of Bohemia in their breeding efforts. The abbot of the monastery was very interested in that question, and he engaged Mendel to look into it. He designed a greenhouse to build for Mendel. He never did build it, but the notion of the little old monk who one day gets this bright idea and goes out in the garden and fools around, that's just wrong. That's not the way it was at all.

DC: So you would very much agree with the people in science studies who have been saying, during the last couple of generations, that we need to recognize science as a social institution?

RL: The history of science is a history of the social structures of science and what surrounds them. Absolutely.

DC: Do you feel yourself to be included?

RL: Oh, sure. I mean, the thing that gave me a certain reputation in my field was solving this problem of how to know what genetic variation was like. That was a problem set to me by my professor. I didn't invent it. That was the problem that he worked on his whole life. I'm just an epigone.

Science Is Part of
the Social Structure

RUTH HUBBARD

Clearly, science wasn't all good. It wasn't enough to say, I'm answering
interesting questions. It wasn't good enough to say, I'm answering
questions that aren't doing anybody any harm, because there were
clearly questions being asked and answered that were doing plenty of
harm: weapons, chemistry, gas, plastering the scenery with defoliants.
It just became imperative not to close my eyes to the fact that science
is part of the social structure.

— Ruth Hubbard

Ruth Hubbard spent the first almost twenty years of her scientific
life at a lab bench investigating the biochemistry of vision. Her late
husband, George Wald, who directed the research, won a Nobel Prize
for the discoveries their team made about how the eye works. In the
1960s, during the Vietnam War, her horizons expanded to include the
politics of science. She took a leading part in the emerging feminist
critique of the situation of women in science and authored one of the
movement's most widely read texts, *The Politics of Women's Biology.*
Later, she became a fierce opponent of the direction biology was taking
in developing new genetic and reproductive technologies that amounted,

in her view, to an experiment on human beings. With her son, Elijah Wald, she wrote *Exploding the Gene Myth.*

Ruth Hubbard was the first woman to hold a tenured professorship in biology at Harvard, and today she is professor emerita. I interviewed her in her office there, and she told me some of her story. She was born in Vienna, where both her parents were doctors. When she was fourteen, her family fled to the United States following the Nazi annexation of Austria in 1938. There, Hubbard attended high school and university, entering Radcliffe the same year the United States entered World War II. Her career in biology began with a job in the lab of one of her professors, and later her husband, George Wald.

RUTH HUBBARD

It was a really fun period and very exciting. It was an easy way to get hooked on science, because we designed all our own instruments. We made everything that we needed — there was a war on, you know — and George had gone to a technical high school in Brooklyn and could use tools. We just did everything ourselves. Every once in a while, we had to give something to somebody to fix, but it was exciting. It was good.

DAVID CAYLEY

What was going on in this lab was research on the biochemical reactions that enable the eye to see, work that would eventually win George Wald a Nobel Prize in 1967. When Hubbard joined the lab towards the end of the war, they were following up on a discovery that Wald had made just after he completed his PhD in 1932.

RH: He went on fellowship to Germany, and he essentially found the role of vitamin A in vision. There had been some thought that vitamin A must have something to do with vision, because there was this business about night blindness if you were vitamin A deficient. George found that vitamin A was somehow involved with the pigment in the eye, in the retina of the eye, that absorbed the light, and that the light absorption liberated vitamin A. He also found another compound related to vitamin A, which he called retinene because it was in the retina. He described it as an intermediate between the visual pigment, which is rhodopsin, or visual purple, and vitamin A, which was the product of bleaching. Bleaching first produces this yellow thing, which he called retinene, and then vitamin A. They are carotenoids — the molecular structure of both retinene and vitamin A is related to beta-carotene, the thing that makes carrots yellow or orange. I began to go to work on how retinene, this yellow thing, got turned into vitamin A. Essentially, I was doing enzyme chemistry. What we opened up in relatively short order at that point was the actual chemistry of this cycle that George had worked out: you have rhodopsin, which builds up in the dark. When you shine in a light, you get a retinene. It gets turned to vitamin A. When you go back into the dark, the vitamin A or the retinene can go back to rhodopsin. We tried to work out the chemistry of this, and did. I did that for quite a long time, into the late fifties — early sixties.

I was really not particularly politically aware at this time. I mean, I was very interested in other things going on in the world, but I was not at all thinking about where the questions of science come from or how the situation in the world affects science. I was perfectly comfortable in the internalist paradigm in which you ask a question and you try and find an answer. If you're lucky, you find an answer, and that brings up the next question, and so on and so on. You don't really look very far to the left or right to ask why some questions are more important than other questions, or why you would want to do this rather than that, or maybe not do this rather than that. Even with things like the atom bomb — obviously I had opinions about that, and there were certain kinds of science I would not have wanted to be doing, and I knew people who dropped out of physics and went into sociology or history of science or something after the bombs had been dropped, but such matters didn't loom that large for me, really.

DC: What Hubbard calls the internalist paradigm holds that science follows its own agenda, asking and answering questions that are internal to the various sciences. It's an accurate enough description of the work that was going on in her lab — no obvious social or political agenda was driving that work — but, as time went on, it came to seem less and less satisfactory to her as a general description of what animates the sciences. What changed her mind was the emergence of the women's movement and the Vietnam War.

RH: The Vietnam War raised the question of whether there might be other problems out there which might be bigger problems. Why am I doing this? Also, science was involved in the war, so clearly science wasn't all good. You know, it wasn't good enough to say, I'm answering interesting questions, and it wasn't good enough to say, I'm answering questions that aren't doing anybody any harm, because there were clearly questions being asked and answered that were doing plenty of harm: weapons, chemistry, gas, plastering the scenery with defoliants. It just became imperative not to close my eyes to the fact that science is part of the social structure and that doing science just because it's really interesting and fun and saying, well, as long as I do good science and I'm answering interesting questions, that's enough — maybe it wasn't enough. I became quite restless about that.

Of course, the women's movement was there in a much more direct way, because it really challenged me to think about — not how we thought about vision, because that was not a particularly gendered topic — but how we thought about biology, how we thought about women's biology, how we thought about evolution. The women's movement had been showing that literature, history, and psychology are all affected by the fact that for social reasons, for societal reasons, the big questions have been asked by men and not by women, and men asked certain kinds of questions, which were the questions that were of most interest to them. When women started looking at this situation more critically, they began to point out that there were certain questions that hadn't been asked, or if they were being asked, they were being asked in strange ways so as to give strange answers that didn't really correspond to the experiences of women. It seemed clear that science — I mean biology, the science I was dealing with — wasn't that different, so that observation must be true in biology, too.

I just decided to take a sabbatical. I didn't have a job — that was another interesting thing that I suddenly became aware of: that I was a research associate and lecturer, whereas the men who were my contemporaries were either on the ladder or already had got professorships.

How come? For a little while, I was naive enough — and I understand how I got there — to think that that was really rather nice, that I didn't have to do a lot of the things that the men who were on the ladder had to do. I didn't have to go to boring meetings. I could pick and choose what I wanted to teach. I could be a lecturer and do this or that. What's so bad about that? Except after a while I got a little older and my kids were growing up, and I realized that I had, you know, no security except through who I was married to and that I wasn't really looking at the whole scene.

At that point, two things happened. One was that two of the younger women who had just come to Harvard and were assistant professors — because Harvard had noticed that there are women who could be assistant professors — called a meeting of all the women with Corporation appointments, which all of us had. I found myself in this room full of women, many of them my age — so, you know, not in their twenties and thirties, but in their forties and maybe early fifties — who'd been here for a long time, who'd done a lot of good work, and who hadn't any jobs, except for these non-jobs that we had. That was an eye-opener for a lot of us. We'd just been naively thinking, oh well, Harvard lets us work here. Isn't that nice of them? We get our grant money and we're doing what we want to do and that's just great. That was one side of things.

The other side, which was more interesting — though not as practical — was that I realized that if it is really true that science — let's say biology, since I'm a biologist — like literature and history and psychology and all those things, is affected by the fact that it's been done largely by men, then how am I going to figure that one out? How am I going to test that? It occurred to me that the basic work in biology, after all, was *The Origin of Species*, and I hadn't read *The Origin of Species* since I'd studied for my generals. How about having a look at *The Origin of Species* and seeing whether that shows male bias? I couldn't have picked a better book. And not only male bias, but just social bias. First, there's the animal kingdom. How come a kingdom? Then, it's operating in a world of scarce resources, so that all the organisms in there, in any one species and between species, are in competition with each other. Where does that come from? Is that obvious? Is that the only thing you can see when you look at the world of nature? Then, you know, you find that there was a Russian prince, Peter Kropotkin, who tells the story just the opposite way. Isn't that interesting? We've got this Brit sitting in England, and he lives in a world of scarce resources where there is competition of all against all, and then we've got this Russian anarchist prince living in a world of cooperation, and that's what makes things work. And so on.

Of course, in Darwin you also get sexual selection. You get this marvellous Victorian script working itself out in which the males all compete for access to fertile females, while the females all sit around at the sidelines, like at a Victorian dance. They try to be choosy about the best man/male they can pick in order to advance their own situation in their world. Well, you know, that sort of changed things for me, and I at that point decided to teach a course on biology and women's issues. It was a seminar course, but I decided to start it off by doing a thing on evolution and Darwin, out of which I wrote this essay called, "Have Only Men Evolved?" It got reprinted in a lot of places, and it started us on actually putting out the first book of that sort. We called it *Women Look at Biology Looking at Women*. That was interesting because at just about the same time as we came out with this book with a weird title, *Women Look at Biology Looking at Women*, there was a collective in England that came out with a book called *Alice Through the Microscope*, which essentially raised the same set of questions. That's the beginning of that story.

DC: Hubbard's objection to Charles Darwin's story of sexual selection and the universal struggle for life is not that it is wholly untrue, but only that it is made out to be the whole story. She mentions the Russian prince Peter Kropotkin, a generation younger than Darwin, who challenged the Englishman's theory on the grounds that it completely omitted the role of sociability and mutual aid in evolution. Nature, Hubbard says, provides grist for many theories, and what we find is very often what we set out to look for.

RH: If you look at animal models, you can find examples of all kinds of behaviour. There are cases where the females are bigger and the males are smaller, where the females are more competitive and the males are less competitive. You can find it all. But the animal models that are out there, and that the children are learning about in school, are the ones about, you know . . .

DC: . . . the two big bucks banging their heads together.

RH: That's right. And when you find examples where it doesn't work that way, then that becomes the problem. Why *isn't* it that way?, not, why is it that way? We just read our social arrangements onto the world of animals and plants, starting at the simplest level of having kingdoms and the lion as the king of the beasts and things like that. Then it turns out that it's the lioness that does all the doing, and all the lion knows how to do is growl louder.

DC: Social structure, in Hubbard's view, affects science in at least two main ways. First, it colours what we perceive in nature — the process she calls reading our social arrangements onto the world. Second, it influences the questions that science takes up in the first place — what it considers worth doing — a problem with which she was particularly concerned in biology.

RH: From the point of view of the effect political and social theory have on the science that gets done, I got very involved, on the one hand, in eugenics, and, on the other hand, in recombinant DNA, which I found offensive, not in and of itself, but offensive in the way it was being pursued. Genes were being exchanged that had never been exchanged in nature before, or maybe exchanged very rarely. It was being done in colonies and in laboratories on a large scale, and people were acting as if they could foretell the outcome and predict whether it would be beneficial or not.

DC: Recombinant DNA, first achieved in 1973, involves the insertion of DNA from one organism into the genome of another. Hubbard's worry was that it was being pursued recklessly, and, as other genetic technologies emerged, her concern grew. She began to notice, for example, how genetic screening of pregnant women was undermining the aims of a movement with which she was identified, the women's health movement.

RH: With the women's health movement, as it came up in the sixties and into the early seventies, the cry was, give us our bodies back. What did that mean? Stop the medicalization of women's health. The medicalization of women's health meant that, in the course of the nineteenth century, as part of finding their niche in society, doctors had displaced midwives and non-medical school and university trained healers, of which there were plenty, by coming up with theories and technologies that gave them an edge on the midwives and healers who assisted in that most fundamental of human capacities — to wit, childbearing. Doctors appropriated childbearing, saying, we know the right way for you to do this, girls. In the course of the 1960s and seventies, and even beginning in the fifties, there was a real effort to take our bodies back and to get the doctors out of the birthing room, if at all possible. Well, at that point, you began to get the development of all kinds of technologies and theories which tried to predict, through various techniques, the health of embryos and fetuses before they were born. That then allowed the medical profession to come back into this area, because how would women allow themselves to have a home birth without any of these predictions when all those dreadful things that doctors could predict might go wrong? You got, first, a takeover — again — and you also got this eugenic pressure, that is, that if you aren't careful the species is going to deteriorate, because you're going to have all these babies that could be avoided and prevented.

What I'm saying is that, at this point, the interplay of medicine and society, of science and society, became very acute and very obvious. You had great inventions like in vitro fertilization, and then you immediately get people like me coming along and saying, how can you do this? How can you say that you're providing a benefit when you're going through manipulations, the effects of which you have no way of knowing? By that I mean using hormones on women to produce multiple ovulations. You have no idea what that's going to produce. You then do the fertilization in vitro — you manipulate this embryo, and you put it back in. You hope that you have done the right thing to prepare the uterus in the right way and that you have the gestation process under proper control, though it hasn't been under its own control because you have psyched it all out. Then, of course, if you want to get to the ultimate of the things that are happening now, you have this huge increase in multiple births. If someone has twins, obviously you welcome your twins, but to go in for a technology that increases the incidence of multiple births — twins, triplets, quadruplets — and do that with the goal of improving health — it doesn't make any sense whatsoever. In all of these matters, the thing that I haven't said yet, but that to me is at the very top, whether it's recombinant DNA or birth manipulations, is that it's all for profit. That's one reason why America is way ahead, because, of course, in the other Western countries you have one or another form of national health system or national health insurance, and in those countries, the fact that this is so goddamn expensive has to limit its use and limit the way in which it gets hyped up.

DC: In the years since the first test-tube baby was born, more than 100,000 children conceived outside the womb have been born in the United States. A British study, done in the late 1990s, reported that of the 7,000 in vitro babies born in Britain in the year of the study almost half were multiples. New technologies have been adopted so quickly and so uncritically, Hubbard says, that they amount to a gigantic experiment on people.

RH: Just do remember that the first in vitro baby was born in 1979. That's nothing. She isn't even thirty, and they didn't use all the hormones that are being used now. They had had no luck whatever getting any babies, so they decided, maybe we'd better not use hormones, and then they did this incredible thing of catching one egg being ovulated and working on it. So what do we know? We know nothing. And it's a terribly difficult thing to be talking about, since we're in a world of lots of in vitro babies.

DC: In vitro fertilization is one of several new techniques that have remedicalized women's health. Another is the genetic testing which now makes women aware of every possible eventuality of their pregnancies: a one-in-so-many chance of this, a one-in-so-many chance of that, and so on. The trouble is, Hubbard says, that these probabilities are just that — probabilities — with no one any the wiser about her actual pregnancy.

RH: I think it's infinitely cruel to women. I think it's the ultimate cruelty to give women information that is meaningless and make them believe that they have to make judgments on that basis. You see, I think one of the things that is so ghastly about it is — I don't know how you intuit yourself umpty-ump years ago — if those tests had been available, would you and your family or I and mine have had either the sense or the strength to say, go away? Or would we have somehow felt a little under compulsion? If it's a first child, okay, but if it's a second child or a third child, what are you doing to the rest of your family? What if — and so on. Now all I know is: you just can't begin. If you begin, if you say yes, then everything else follows. How do you know where to stop? The only thing you can do is say no, just go away, will you? Just leave me alone. Certainly, in my own life, that's what I have learned. I have these conversations all the time. They're not about childbearing anymore, and I wouldn't dare to advise a younger woman unless somebody really came to me with a question and really wanted my opinion. I just wouldn't dare. But with my own friends and older women, even to a little bit younger women, but not much, all I can say is, no, I don't go for an annual checkup. No, I don't know what my cholesterol is. No, I don't know — I mean, all the things everybody else knows about themselves, I don't know. What would I do if I did know it? What does it mean?

DC: What does it mean, Hubbard asks, to know what might happen to you or what might be wrong with your unborn child? Nobody knows what is actually the case. One knows only the risk. In a large enough population of a certain kind, X can be expected to occur with a certain frequency. What are *you* going to do? This is the question in which she has trouble discerning the meaning.

RH: What it means that is not statistical I simply don't understand. You know, you get smart people who have these tests, and they say, oh, but this means that there's a ninety per cent chance of this, that, and the other. Yes, and? Ten per cent chance of not. Yes, that's bigger than if there were a thirty per cent chance, but it's very rare that it's a ninety per cent chance. It's much more usual that it's a ten per cent or fifteen per cent chance that something might happen. But, you know, then people can't live with that, so they go to the next step.

DC: The next step, when it's a question of prenatal screening, can only be an abortion, since there are as yet no intra-uterine therapies. Can such decisions be made less of a shot in the dark in the future? Only to a certain extent. There is the promise that medicine will one day be tailored to one's individual genome, but how does one interpret an individual genome?

RH: They may be able to tell you what your DNA sequence is, but what it means is still going to be probabilistic. There's no way it can be anything else. It's cheating to say, oh, we aren't going do it just on the basis of statistics, we're going to do it on the basis of what your own DNA profile is. How are you going to know how to interpret my DNA profile? Just by looking at my DNA profile? That isn't going to tell you a damn thing. You're going to have to look at a lot of different DNA profiles and see how they stack up. We're right back in the probabilities, aren't we?

DC: Hubbard has many concerns about genetic screening technologies. They impose impossible choices. They make women dependent on the experts who administer and interpret the tests. But overriding all others is her concern that gene technology is being developed, not in the public interest, but for private profit.

RH: I think that the big point in all of this, really, is that this is all being done for profit, and there's money to be made at every step. The point is that, once you have a machine that gives you these statistics or predictions or whatever, you amortize the cost of having bought it by using it. Once you have to have one of them in every doctor's office, or in every other doctor's office, then it's in the economic interest, and not just interest but necessity, of that doctor or that health clinic or whatever to persuade you, and you, and you, and you, and you to let them use it, because that's how they're going to pay for it.

DC: The word "eugenics" refers to the idea that the genetic fitness of the human race, or some part of it, can and should be improved. The term was coined by Charles Darwin's cousin Francis Galton in 1883. The Second International Eugenics Conference in 1921 defined it as the "self-direction of human evolution," and this definition is very much in the spirit of Galton's conception of his new science as a kind of applied Darwinism. In the English-speaking world, eugenics claimed many prominent adherents, including H.G. Wells, George Bernard Shaw, and Margaret Sanger, and resulted, in many jurisdictions, in widespread compulsory sterilization of those thought unfit to reproduce. Nazi Germany defined itself in eugenic terms — National Socialism, one of the party leaders, Rudolf Hess said, was "applied biology." But the horrors that resulted in Germany for a time completely discredited the whole idea, and eugenics, during the period after World War II, was generally treated as part of a less enlightened past. Today, gene technology has created a new form of eugenics, though only opponents apply this highly charged label.

Ruth Hubbard has been one of the most outspoken of these opponents. From the very beginning of molecular biology's effort to understand human beings as products of their genes, Hubbard saw that this undertaking was bound to lead to questions about "who should and should not inhabit the world," as she put the question in her book *The Politics of Women's Biology*. It remains to be seen what will happen as genetic information accumulates and researchers begin to discern genetic tendencies to violence, addiction, or other unwanted behaviours. For now, leaving aside sex selection, contemporary eugenics generally uses genetic screening to prevent the birth of babies who may suffer disease or disability. Behind it stands what Hubbard calls "the gene myth," a phrase she and her son Elijah Wald used in a book they wrote together called *Exploding the Gene Myth*.

RH: If you know that somebody is going to have whatever it is the gene says the person will (or may) have, that does not tell you what kind of a person he or she is going to be. It does not. The fact that it has blue eyes doesn't tell you what kind of a child you have, unless you happen to grow up in Nazi Germany. Life is very complicated, and it is part of our job to see to it that it remains complicated, that we don't go for simplistic questions and don't go for simplistic answers.

DC: The gene myth that you and your son wrote about in *Exploding the Gene Myth* — what did you mean by the gene myth?

RH: That genes are predictive. That genes are predictive at the individual level, that by knowing your genes you will be able to know your future in specific instances. What we were arguing there is that even in the most reliable instances, you still don't know it for the individual. In most instances, you really just know the statistics.

DC: Among the reliable instances that Hubbard speaks of is Tay-Sachs disease. It's a fatal condition that occurs largely among Jews of Eastern European origin. They suffer it at a rate of one in 3,600, while its rate in the general population is one in 100,000. An accurate test has been available since the early 1970s to detect carriers of the relevant gene. This is the best case for genetic screening.

RH: If your fetus has a double dose of the gene that mediates Tay-Sachs, then the chances are pretty darn good, or bad, that the child will have a disease that he or she will die from within a few years. If you want to look the genie in the eye, then yes, you do need to test both parents. If both parents are positive, then you can say that — I mean, it's a recessive gene — that there's one chance in four for every child to have the disease. If you feel you can't live with that . . .

DC: That's the strongest known case, the most certainty one can have in this area?

RH: One in four is the most certainty you can have in this area, and it's one of the strongest cases because it is one of these rare conditions that is almost invariably — or maybe you can even take out the almost — fatal in a short time, you know, by age two or three or four. However, I have heard families who have had this experience argue, and families with children with other conditions have also argued, yes, and that, too, is part of the human condition, and do I feel that this was an all-bad experience for our family? No, I do not. It depends on how you look at the world. But let's talk about Down syndrome, which is, after all, the one that probably most abortions are done for. Certainly nowadays, with improved or different ways of dealing with children who have Down syndrome, you have a lot of families saying, you know, this is our child. Then you also have families saying, this is our child, and we're going to have another one. We would really like the other one not to have Down syndrome, because we would like to have the experience of having the other kind of child. It does depend so much on what you think life is about and what you think the world is about, what you think health and illness are about, and life and death.

DC: Ruth Hubbard began our interview by recalling her days in the laboratory studying the chemistry of the eye in the 1940s and 1950s. It was science pretty much for its own sake. She and her colleagues were just trying to understand how something works. There was no pot of gold at the end of the rainbow, no lucrative technological spinoff in view. A lot has changed in the intervening years. Biology has become big business, and a lot of solacing ideas about the purity and high-mindedness of science have also fallen by the wayside. There are exceptions, of course, but science generally, she has concluded, is no better than the society it belongs to.

RH: Science is part of the culture. We have come around to the fact that science is another money-making enterprise, and there are lots of new ways of making money in it. Furthermore, it is an enterprise in which the prize goes to those with the most money. It's no more pure than anything else. It's probably purer to go out there selling potatoes. Look, in a way, we were living in Never-Never Land, because things have

only changed for people like me, who grew up thinking of biology as a
poor science, economically poor. The chemists have lived in this world
forever. Forever, I mean, since the time of Haber, let's say, and World
War I, when they really hit it big with companies like I.G. Farben. The
physicists hit it big in World War II with all the toys that they produced,
including the bomb. Biologists had the privilege of living in the age of
innocence for a much longer time, so that when I started there wasn't
that much money there. I don't know — I'm going to be fantasizing —
but I bet you that George's first grant was for $500 or $1,000. Then
they got to be $10,000. Oh, my God. I remember when we published
something or gave a talk or something which showed that when we
used a vitamin A that had come from fish liver oils, we got a lot of
rhodopsin synthesis, but when we used crystalline vitamin A that came
from some pharmaceutical house, we got hardly any. Boy, the next day
or the next week there was the phone call about fish liver oil. They took
us out to lunch, and so on, and offered us money for doing this, that,
and the other, and proving that fish liver oil was better for your vision
than crystalline vitamin A. We laughed, you know, we weren't going to
go into that. We then did, in fact, try to figure out what the difference
was, and that led to something interesting. It turned out that vitamin
A comes in different shapes, and only one shape is the right one for
making visual pigment. It gave us five or ten years' worth of interesting
work. But we were not in the big time in the same way that biology got
to be in the big time with recombinant DNA. You could tell immediately
that that was going to happen, because all the scientists suddenly were
queuing up for their monies.

DC: Hubbard has stuck to the conclusion she first came to in the 1960s
when she began to think about the role of science in the high-tech war
that was then being waged in Vietnam: science must inevitably be a
creature of its society.

RH: Unfortunately, it's the social system as a whole that matters. It's
the politics. It's the social system that matters. Being a scientist doesn't
really let you off. You can, I guess, buy yourself out, but you have to do
it very consciously and know why you're doing it and how you're doing

it. So I think as long as the current ethics run the show — and I don't like the word ethics — it's politics that I really mean — the politics has to come first. Unless the politics comes first, if you're for sale, you're going to be bought.

Science and Myth

MARY MIDGLEY

What I feel we have to do now is to alter our way of living as fast as possible. But we can't alter it without first altering our thoughts.
— Mary Midgley

Mary Midgley is a British philosopher who has put our relationship to the earth and to other animals at the heart of her work. She began her writing career relatively late in life, after first raising a family, but thirty years after her first book *Beast and Man* appeared, and now well into her ninth decade, she is still writing and still wrestling with the great questions of our time. The book that drew me to her work was her 1989-1990 Gifford Lectures, published in 1992 as *Science as Salvation: A Modern Myth and Its Meaning*. It deals with the sometimes heroic, sometimes prophetic role science has assigned itself in modern civilization and with the stories scientists have told about their enterprise. I interviewed her at her home in Newcastle in the northeast of England, and I began by asking her about the sense in which she had spoken of science as a modern myth

MARY MIDGLEY

Now, I obviously don't mean "lie." One can use the word "myth" simply to mean lie, but that's not an interesting use and it isn't what I'm talking about. I mean an imaginative picture, a drama, a dream which people are fascinated by. All of our thought has to have an imaginative and an emotional side, and the need is simply that we understand that that's so. It has to use metaphors. All science does use metaphors, and it has to have some sort of emotional background, some direction from which it comes.

DAVID CAYLEY

Science takes on this narrative and emotive colouration, Midgley thinks, as soon as it tries to express itself. This is obviously true when a scientist like Stephen Hawking puts on the cloak of prophecy and tells us that a complete and unified physics will one day disclose "the mind of God." But even the scientist who heroically refuses all meaning, sticking to the bare facts in a barren cosmos, is still, in Midgley's view, telling a self-dramatizing story. And this storytelling proclivity goes right back to the origins of modern science in the seventeenth century.

MM: The things that happened in the seventeenth century, I think, have made a tremendous difference, and we are still stuck with certain ways of thinking which were invented then. I mean, one thing that came on in the seventeenth century, and has really only increased since, is this tremendous reverence for science as such and the incredibly high hopes that have been put on science as providing the answer to all our questions.

It's quite interesting why seventeenth-century people got so keen on this, so keen to think there is a simple structure to the universe. If only we can find that, then everything will come out right. Newton's clockwork universe, which did seem beautifully simple, was just what people needed. What strikes me now, thinking about it, is the appalling confusion of the seventeenth century, particularly the Wars of Religion, you see, which were both really messing up people's lives directly and also undermining what they felt was the foundation on which they stood — because if there's doubt about your religion, then what isn't there doubt about? That sense of deep confusion was the source out of which this obsession with total order and total simplicity arose.

DC: This quest for total order, as the natural philosophers of the seventeenth century understood it, was an explicitly masculine undertaking — another of Mary Midgley's examples of how stories colour science.

MM: One unfortunate fact about those seventeenth-century theorists — and they would be staggered if they knew how awful it looks now, because it seemed quite obvious to them — was their conviction that science was a masculine activity and that it was terribly important that it should be so. What this masculine activity was doing was hunting a female called nature — looking for her everywhere, finding her secrets, and not putting up with any nonsense — you know, digging her out of her hiding places and piercing her in order to provide the final truth. I mean, they used gender imagery of an extremely crude kind, and they didn't know how crude it was. What they were doing — and this is worth thinking about — is that they were reacting against another school of scientists at the time who liked the idea of Mother Nature and of the soul of the earth — *anima mundi*, wasn't it? They were prepared to say that gravitation, for instance, was the love which the various objects, physical objects, felt for the source from which they came — their mother — and that nature was a bountiful mother providing all sorts of things that we don't fully understand. Now, the Baconian and Royal Society people thought that this was terrible, superstitious stuff. We can't have any of that; we don't want emotion getting into science, they said, although their own writings are extremely emotional. They only recognized emotion when it was of the tender and affectionate kind. They didn't think of their own aggressive and destructive approach as being emotional. This is a way of talking that still continues. If you complain about the treatment of pigs, people say you're getting emotional, but if someone defends this treatment on the grounds that we want our profits, don't we?, that's not emotion. It was terribly unconscious. That's what I'm saying. They might have used these metaphors and then asked why those particular metaphors were being used — that's always interesting — but they didn't do that.

DC: Scientists have tended not to notice acceptable metaphors while at the same time expressing violent antipathy towards unacceptable figures of speech. This use of imaginative structures that resonate with the spirit of the age has carried on into our own time. An example is Richard Dawkins's *The Selfish Gene*, published in 1976, which quickly became one of the most widely read and widely discussed scientific works of its era.

MM: Here's a drama, isn't it? It's a drama for an individualist age, with a lot of little individuals, right inside and under the surface, each pursuing its own interest only and engaged in endless competition. *The Selfish Gene* caught on, you see, because it is written with great imaginative force. It gives a picture of these dreadful little things, the true immortals, beavering away, conquering each other, and the organisms, including us, in which they work are said to be "lumbering robots" being pushed around by these horrid little creatures. I mean, I'm saying that this was totally satisfactory to an individualistic age because it said, not only are you behaving like this in your economies, but everything in life behaves like this. Like many myths, this had got some truth in it — there is a lot of competition in the living world — but it hadn't got half so much as Dawkins claimed. As people have pointed out lately, you can't have competition at all unless there's a great deal of cooperation for a start. And there *is* a great deal of cooperation in the living world. If one had said, this is a cooperative universe, this would have been just as convincing, but that it should be competitive is what the Thatcherite and Reaganite age really wanted to hear.

So, you see, what I'm saying is that, although these myths usually do enshrine some real scientific facts and things that people need to know, they are also powered by some wish of the age, and one does just have to look out for that. That's all. What often happens, of course, is that one myth can be used to counter another. We don't have to sign up for just one of them. In the seventeenth century, for instance, the idea that

God's purposes are something much bigger than us and we shall never understand them provided quite a healthy balance to this thought that we shall quite soon get the final answer to everything. As people have lost that, they have lost a good sense of reality about the limitations of our own thought.

DC: Richard Dawkins's *The Selfish Gene* belongs, in Mary Midgley's opinion, to a class of scientific works which slide surreptitiously from science into mythmaking and often achieve unusually wide circulation in the process. Another equally prominent example is a book called *Chance and Necessity*, which first appeared in English in 1971. The author, Nobel laureate Jacques Monod, was a distinguished French geneticist.

MM: Monod, who wrote a influential book called *Chance and Necessity*, had this myth of the casino. Life's just a casino. It's sheer luck, you see, that we're here at all. Everything's sheer luck. Now, a casino is not a neutral sort of model. It's quite a powerful emotional model, isn't it? You feel you've been helplessly thrown into this gambling hell. It's a highly social way of thinking. It's not just black and white. It's highly colourful, in the same way *The Selfish Gene* is highly colourful. And, of course, the colour that Monod was confirming was the colour that was set up by social Darwinism — "nature red in tooth and claw."

DC: Midgley regards Jacques Monod's *Chance and Necessity* as, in her phrase, "an existentialist tract." But it presents itself as a work of science, and this is the basis of her objection. She thinks that books like *Chance and Necessity* are guilty of both a certain arrogance and a certain naivety in failing to acknowledge their philosophical and ideological biases.

MM: It would be a very good thing if more scientists were more aware of these big philosophical issues, more aware that they have got a bias towards seeing the world one way or another and not thinking that they are, as it were, directly reporting the world. The delusion that one's own metaphysic is a part of science is pretty widespread. What's bothering me is not that science itself is divisive but that scientists, by the way they get educated, tend, on the whole, to have far too specialized a view and to be not much aware of these larger issues. Now, this is not half so true of scientists who have been educated on the continent of Europe — except Monod. They do tend to do some philosophy and some history, as well as their science courses. It's this splitting of science from the rest of culture that I think is so unfortunate. It's particularly bad in this country — this narrow education of scientists. If you aren't given that more general perspective, the temptation to think that you have got the answer to everything is very great. So they go about pontificating about whether God is there or not, and so forth, and I don't think they should.

DC: Midgley dislikes what she calls pontification by scientists. Jacques Monod, in her opinion, has no scientific warrant for asserting that the universe is a giant casino, any more than Richard Dawkins has for claiming that human beings are robots controlled by their genes, but these statements are presented by their authors as if they were part of science. This conflation of science and ideology, in Midgley's view, often intensifies popular reaction against science. The current culture war over evolution in the United States is, for her, an example. There is, obviously, a real disagreement between science and a literal reading of the Bible, but she thinks that the problem has been very much aggravated by the way evolution has been presented. This goes all the way back to the nineteenth century, when Darwinism was conveyed to the American public in the form of what was called Social Darwinism — "nature red in tooth and claw," "survival of the fittest," and so on.

MM: Herbert Spencer, who invented Social Darwinism, went to the States and preached it all over the place in the 1880s to such effect that he outsold every other philosopher in the United States for the next ten years. So Social Darwinism got into the water supply, and that's what I think Christian people who are shocked by a Darwinian or scientific position are thinking — "nature red in tooth and claw" and a casino. The irresponsible use of those myths has had a terrific effect, and the scientists often delude themselves that they are only giving people science when in fact they're giving them moral and political prejudices of their own.

DC: "Every thought system," Mary Midgley has written, "has at its core a guiding myth, an imaginative vision, which expresses its appeal to the deepest needs of our nature." This being her view, she doesn't think that science will ever outgrow myth. On this score, she hopes only that scientists will become more conscious and more careful about the stories they choose to tell. But she does think that science needs a new myth, and her candidate is her countryman James Lovelock's Gaia theory. In fact, she has written a short book on it called *Gaia: The Next Big Idea*. The Gaia theory holds that Earth's biosphere as a whole is a self-regulating system — that the earth's unstable atmosphere, for example, is regulated and maintained by an ensemble of geological and biological processes occurring at the surface of the planet — and it makes the ancient Greek earth mother Gaia the emblem of this wholeness. One of the things Midgley likes about this theory is the way it unifies the disparate and sometimes antagonistic sciences that have studied different aspects of the life of the earth. Another is the way in which it revives the idea of a living and creative nature.

MM: The whole biosphere has been doing terribly clever and complicated things which we would have had to have been very clever to invent if we'd had to invent them, which we didn't, did we? Matter is not this inert, silly stuff. It has a great deal of potential in it for knowing the right thing to do and going on doing it. The genesis of form comes up out of matter itself and gets more and more complicated. All these proteins and things build themselves. It's not a casino, you see, as Monod suggested. It's not a situation in which absolutely anything might happen, because what will happen will be what suits the kind of molecules you've got already. The more complex they get, the more there are particular paths in which they will go, and these lead to certain forms.

If you like to look at it theologically, the immanent god in these things is making this happen. You don't have to have an artificer god with a hammer making it happen from outside.

DC: The founding myths of modern science imagined humanity as distinct from nature, and nature as brute mechanism without inherent purpose or direction. The new myth that Midgley thinks science is developing, and badly needs, emphasizes the creative spontaneity of nature and locates humanity within this burgeoning and evolving universe.

MM: There's some sort of general tendency towards the more complex, isn't there? And at some point it produces consciousness. Now, this seems to me a perfectly continuous sort of thing, whereas Descartes said that consciousness is something quite different that is put in from outside. We don't have to think like that, do we, and I've never been inclined to think like that. When people talk about the Anthropic Principle — that the whole point of the cosmos is to produce us — I find it rather odd. Why not the Giraffic Principle or the Elephantine Principle or the Black Beetle Principle? All these other creatures, as much as us, need this particular cosmos to be going on in. They couldn't live without it. That there is a movement towards life in general seems to me perfectly sensible, and then, when you get that, of course, there's a movement into divergent forms of life because there are different niches, places where you can live. The point has to be that matter has got certain ways in which it does naturally go if it's given a chance.

DC: Humanity, for Midgley, is continuous with the rest of nature, kin to everything that has emerged and will continue to emerge from the womb of matter. However, a very different orientation is built into the institutions of the modern world. To her, how we act depends ultimately on how we think and imagine — which is why, for her, philosophy is no idle pursuit — and the ecological crisis which we are entering is the product of the way the modern world has thought.

MM: Since about the seventeenth century, we have taken a much more exploitative, confident attitude, haven't we, towards the physical world than people did before. Certainly they exploited what they could easily find, but when they found things were going wrong, they rather quickly said, oh dear, we've made a mess, it isn't God's will, or something like that. They often didn't do it because they were frightened in the first place. This enormous confidence with which Western people have taken over everything since the seventeenth century is a matter of how they think, isn't it? It's a result of them thinking differently about all this stuff, of their believing themselves to be much more important and much more authorized to make enormous changes than they previously did. In Greek tragedies, if you take on some enormous initiative, you're rather likely to come to grief, and that was still so in Christian thought, although the way in which it happened was a bit different. It's been this great experiment, hasn't it, since the seventeenth century. We'll try just doing all these things and see what happens. What I feel we have to do now — and I do expect things to be pretty bad — is to alter our way of living as fast as possible. But we can't alter it without first altering our thoughts, and particularly our thoughts of what we ourselves are, can we? The Enlightenment idea — now we've got rid of God, we can just take over his position — doesn't seem to me to be working.

An Anthropology of Science

ALLAN YOUNG

We don't have some kind of big carpet, and underneath that carpet is reality, and we're gradually unrolling the carpet, and we're discovering a reality, and this reality is a reality that exists before we have attempted to encounter it. Rather, it is a product of that encounter.

— Allan Young

Post-traumatic stress disorder, or PTSD, is a disease that first appeared in the *Diagnostic and Statistical Manual of Mental Disorders* in 1980. Initially, it was a diagnosis applied exclusively to Vietnam veterans; today, it is part of the vocabulary of everyday life. How this happened is the subject of Allan Young's *The Harmony of Illusions: Inventing Post-Traumatic Stress Disorder.* Young is a professor of anthropology in the Department of Social Studies of Medicine at McGill University. His book traces the idea of traumatic memory from the 1860s, when a British surgeon first described the lingering after-effects of railway accidents, to our own time, when the National Institute of Mental Health in the US estimates that every year 7.7 million Americans suffer from PTSD. More than that, it asks how a scientific object, like a psychiatric diagnosis, comes into existence in the first place.

Allan Young speaks in his subtitle of the "invention" of post-traumatic stress disorder. His use of this word makes a bold, and potentially controversial, claim, but this claim is not that there is anything unreal about PTSD. Young states unequivocally at the beginning of his book that PTSD and the suffering associated with it are real. What he means by "invention" is that post-traumatic stress disorder is not a natural object that, at a certain moment, was discovered, named, and described. Frightening events must always have left painful impressions in the mind. But post-traumatic stress disorder is much more than just a recognition of this commonplace fact. It is a social and scientific construction, which, among other things, establishes a certain theory of memory, unifies disparate symptoms under one heading, confers benefits and social status on its sufferers, and empowers those who study, treat, and certify the disease. In all these ways, PTSD can properly be described as an invention, an act that constitutes its object rather than simply describing it.

Young approaches this object as an anthropologist, someone who adopts the stance of an outsider unprejudiced by the taken-for-granted interests and assumptions of the field he is studying. It was about his formation as an anthropologist that we spoke first when I interviewed him in his office at McGill. He told me that one of his greatest influences was a book that he regards as the beginning and still the summit of medical anthropology: *Witchcraft, Oracles and Magic Among the Azande*, by Edward Evans-Pritchard, published in 1937. Evans-Pritchard was a British anthropologist who studied the Azande people of the Upper Nile at a time when British colonial policy was forcing them to abandon their scattered settlements and move into European-style villages in the interest, the British said, of eradicating sleeping sickness. As a member of the Colonial Social Science Research Council, Evans-Pritchard was called on to explain the Azande's resistance to this policy.

ALLAN YOUNG

The explanation was the fear of witchcraft, and the Azande account of witchcraft made it sound very similar to the effect of firearms. If you live ten kilometres away from someone who's got a rifle and hates you, you don't have a lot to worry about, even if he occasionally plunks shots in your direction. On the other hand, if you're living next door to him, you've got a lot to worry about. The Azande had a very similar idea. This fear of witchcraft, of course, just played into the conventional wisdom about people living in tribal societies as being slaves to custom and having these very strange sorts of ideas. But Evans-Pritchard was with them for a while, and he came up with this very interesting formulation — a puzzle, really — and it is that the Azande, first of all, are totally rational. They are also entirely empirical. In other words, they want to see proofs, and empirical proofs, for whatever they believe. Because they are empirical, the Azande can recognize contradictions in their beliefs and the world. Moreover, the Azande are progressive: they want to improve their lives; they're not slaves to custom. So here you have four propositions. Then the fifth proposition is that they believe in witches, and the sixth proposition is that there are no witches in Azande land. The puzzle is, how do you reconcile the first set of propositions about rationality, about being empirical, with the second set of propositions — that they believe in witches and that there are no witches in their territory. The rest of this mammoth book, beautifully written, is an explanation that reconciles the two of them around a way of understanding how knowledge is produced in the world.

DAVID CAYLEY

So how do an empirically minded people come to believe in witchcraft? Evans-Pritchard's answer is that belief in witchcraft is a product of what Young calls an "epistemic culture." The term refers to what any group of people know in common, *epistēmē* being the ancient Greek word for knowledge. Young illustrates this side of Evans-Pritchard's thought with reference to his second great teacher, Ludwik Fleck. Fleck was a Polish-born medical doctor and microbiologist who survived Auschwitz and died in Israel in 1961, and his thought complements the work of Evans-Pritchard.

AY: For me, Fleck stands next to Evans-Pritchard, not only in my own intellectual development, but, I think, in our understanding of science. Fleck was a Polish Jew, and in 1935 — more or less the same time as *Witchcraft, Oracles and Magic* was published — he published a book, now a very, very famous book, called *Genesis and Development of a Scientific Fact*. In the book, he makes an extremely bold claim, as a scientist, that scientific facts are produced. Another way of saying this is that when people work together and collaborate, they create what you might call an epistemic culture. Fleck used the German word, *Denk-Kollektiv*, the "thought collective." He says that what is going on is not something that is in the mind of a single person but in the collectivity that is doing the research, in the apparatus that they're using, and so on and so forth. These are ideas that we know today.

But, if I can just parallel it very briefly with Evans-Pritchard, Evans-Pritchard makes the argument that witches and witchcraft are a product of an epistemic culture, involving the particular technology that the Azande evolved over a long period of time of oracles of various kinds: poisoning chickens and putting sticks into termite hills and things like that. In the end, his conclusion is, if I can put words into his mouth, that witches and witchcraft were the epistemic product of this culture, and in that sense, they're entirely real. They were real in that people were getting sick, were dying, and so on, but we can say something is real, in this sense, without applying the standards of truth.

Fleck makes a similar point with regard to scientific research, and he uses this beautiful phrase when he describes how facts are produced and stabilized and then circulated amongst communities of scientists. He describes them as being products of "a harmony of illusions." The illusion is that somehow we have penetrated from what is real to us to what in fact exists outside of our efforts to understand the world.

I took the term "harmony of illusions" directly from Fleck, and the argument that I made in the book and continue to believe is that PTSD is in fact real. There's no denying this. There's no denying that witchcraft amongst the Azande is real — it's part of that people's life world. It does not exist simply as ideas and impressions in their minds, but it exists in terms of sickness, in terms of diagnosis, in terms of death, and in terms of all sorts of very important material decisions that people are making. The same thing is true with regard to PTSD. Now, let me clarify something at this point, because one could then say, wait a minute. Are you suggesting then that Azande witchcraft and Azande witches are the same thing as PTSD, since they're both products of what you're calling "epistemic cultures"? The answer is, no, they're both products, but they're products of a profoundly different character. On the other hand, from the point of view of the anthropologist, they are equally open to an anthropological interpretation, an anthropological methodology.

DC: Thought collectives, or epistemic cultures, are by no means all the same, Young says. Reliably connecting a bacterium with a disease, the subject of Ludwik Fleck's study, is not the same thing as discerning the identity of a witch from the entrails of a chicken. Even so, both results are products of a certain technology and a certain collective thought-style, and, in this sense, they can be analyzed with the same ethnographic methods. Applying these methods, however, presents the scientific anthropologist who wants to study his fellow scientists with a set of problems very different from those the anthropologists of yesteryear ever faced.

AY: When I speak to my students, I give them a kind of a history of anthropology and medicine, but it's really my own autobiography. At the very beginning of anthropology, the great task or obstacle that anthropologists had was to draw themselves close enough to the people they were studying. If you look at the great anthropologists of the twentieth century, like Malinowski and Evans-Pritchard, that was their great achievement: being able to draw close. But when you work in your own society, and particularly when you work with people who are, in essence, colleagues rather than informants, whose claims on knowledge are as strong as your own, if not stronger, then the great task becomes, not drawing close, but being able to separate yourself. I would particularly say, not to use pretentious terms, that there's a true ontological insecurity in working with scientists. All of us, I assume, have our sense of what is real in the world and of the authority to which we turn to establish what is real, and for me, it is science and scientists. This is the bedrock of my notion of reality. At the same time, I'm studying it.

That leaves me three options as an anthropologist. Option number one is the Luddite option of saying, these scientists are dehumanizing the world, disenchanting the world, but that's not my view. Option number two is simply to translate what scientists already know into a new language, the language of anthropology, or the language of science journalism, which is fine. I have no problem with that. And then there's number three: to do something different, to do a real ethnography of science and scientific knowledge. That's where the problems begin, and they're not only, say, epistemological problems. They're also moral problems in the following sense: if I had to say, what is anthropology, and what is the goal of the anthropologist, I think one way of answering would be to say that it is the job of the anthropologist to make explicit what's taken for granted by everyone else. It's very difficult to do, even in other cultures. The problem with making explicit what is taken for granted is that, when it is made explicit what the presumptions are, often one's informants get very angry or uncomfortable.

Fortunately, for me, this has happened only a few times, and the area in which I work, the disorder in which I'm most interested, post-traumatic stress disorder, is highly fractious. There are all sorts of arguments taking place within the field and great bitterness on the parts of some people. It is not just my observation, but the observation of other people, people within psychiatry, that, to some extent, there's a religious quality to much of the sensibility of researchers and clinicians working in the field of PTSD. One can understand it, because there are victims and suffering and difficult issues of compensation, but that still makes it a rather difficult field for an anthropologist to be able to draw back from.

DC: Despite these difficulties, Young has now spent more than twenty years studying the field of PTSD. It began with an invitation.

AY: In 1985, the American Congress mandated the creation of a special unit within the Veterans Administration Hospital for developing a treatment program and then an educational program for post-traumatic stress disorder. The disorder had entered psychiatric nosology five years before, in 1980, and there was a lot of political pressure within Congress, particularly within the Senate, to create a unit like this. It had been established for several months, and I was then invited — a rather

complicated story, I won't bore you with it — to visit this hospital and
to visit the in-patient unit — the in-patient unit was the very core of
the facility — and to see if I would like to do some research there as an
anthropologist. This invitation came from people who were familiar
with anthropology, familiar with some of my earlier work, and they
invited me there.

DC: Young conducted field research at this Veterans Administration
hospital in the American Midwest over a three-year period. In 1995, he
published *The Harmony of Illusions*. The book reports on his research,
but it also relates the history of the idea of traumatic memory, a history
that reaches back into the nineteenth century.

AY: If you go all the way back to 1895 and the publication of Freud and
Breuer's studies on hysteria, you'll find that they talk, in that book,
about traumatic hysteria. They say, in a very famous sentence, that the
disorder is based mainly in reminiscences — that is, not in traumatic
events, but in traumatic memories. The events give rise to memories,
but it's the memory that drives the syndrome afterwards. This is Freud
and Breuer talking in 1895, and there were people before them who
talked likewise.

What happens over the next century is that memory remains an
interest and remains a focus, but it is also highly controversial. Through
the period up to 1980, there are fierce debates among psychiatrists
in many countries. During World War I, for example, particularly in
Germany, there were debates over the nature of traumatic memories:
how those memories are formed, how malleable those memories are,
what they're memories of. Are they memories of the past or, as some
German psychiatrists suggested, memories of the future? That is to say,
are they fears of what is going to happen in the future that then become
incorporated as memories of events that have already taken place? These
are very, very subtle accounts, very stimulating, very important. During
this period, memory science is likewise developing, and similar sorts of
questions are being asked about memory. A famous book is published
in 1932 by Frederic Bartlett, whose teacher was one of the great trauma
doctors in World War I, W.H.R. Rivers, and when Bartlett wrote the

book about memory, he gave it an interesting title: not "Memory," but *Remembering*. He makes the point that, when we talk about memory, we're talking about a process. Every time we remember, it *is* an active process of that memory being assembled or re-assembled on each occasion. There's something very dynamic about it. Memories are malleable. These were wonderful, beautiful accounts of memory.

In 1980, all that changes. In 1980, with PTSD, we get an account of traumatic memory that is very different because it is a standard account. It's no longer memory as remembering. It's not longer memory as a process, but, rather, it is memory as an essence that is created on the occasion of a traumatic experience, and, being an essence, it is an object that doesn't change over time. Once it is produced, it remains as it is. It can gradually erode through therapy, but it is not a process. So a variety of metaphors are used. People speak of flashbacks, or of flashbulb memories. There are references to indelible traumatic memories which suggest that memory is something that is very solid. The question then becomes, why? Why this change out of nowhere? In Yiddish, we have the expression *oyf tse loches,* "out of a hole" somewhere, this idea comes. Well, of course, it doesn't come out of a hole. It doesn't just come out of anywhere. We know all the details. I've interviewed all the people but one who were on the *DSM III* committee that made this change, and the answer is quite clear.

DC: The seed of this answer, Young says, lies in one of the names that was commonly given to precursors of post-traumatic stress disorder: compensation neurosis. The term goes back to the nineteenth century, when the traumatic after-effects of railway accidents became an issue for injured and frightened travellers and for railway companies and their insurers. Its use highlights one of the critical features of post-traumatic stress disorder. Unlike illnesses arising from some physiological disturbance or dysfunction, traumatic memories have an external cause. They are a product of events for which someone else is responsible and potentially liable. In 1980, American psychiatry and the American government needed to find a framework for dealing with issues of responsibility and compensation arising from the Vietnam War. Part of the solution was the inclusion of post-traumatic stress disorder in the third edition of the *Diagnostic and Statistical Manual of Mental Disorders*, the *DSM III*, the *DSM* being the American Psychiatric Association's handbook of mental illnesses. But this solution — redefining memory as an essence, a fixed and permanent impression — was very different from the one the British had found after World War I.

AY: After World War I, it's a huge issue, especially in the UK. There are lots and lots of soldiers who have been diagnosed with shell shock. They're pretty miserable after the war. They can't work. Their family relations are poor and so on and so forth. The British have, around 1922, a pretty large inquiry, with all the most important psychiatrists and army doctors of the time. They come up with a kind of a conundrum, and the conundrum is: okay, let's say, since we can't get into their heads, that all the men who claim to have been traumatized really were traumatized. What should we do now? What they want, and are entitled to, is compensation. How shall we pay the compensation? Well, it sounds like a very dull subject — something that only actuaries would be interested in — but from the point of view of science, it's an extremely interesting argument. PTSD is one of those disorders whose history cannot be understood solely in terms of psychiatry but also has to be understood in terms of the law and in terms of forensics. Both of them are determining what we finally call "trauma" and "post-traumatic stress disorder." The argument is, well, we can do two things. We can do what we've tried to do in the past, and that is, we ask men to come in once a year, or once every six months, for a psychiatric examination, and then we have to see what their disability is as a result of shell shock. If their disability is fifty percent, we give them fifty per cent pension. If it's a hundred percent, they get a hundred per cent pension. We just do that, and eventually some of them are going to get better, and then they'll get no pension. The counter-argument is, yes, but we've been doing this for a while, and no one gets better. They don't get better not only because there are some malingerers amongst them who want to continue getting a pension. There are also many in whom this psychological process is not something they are even aware of — they don't know what is going on — and there are real incentives not to improve. What the paying of these pensions will do is simply create a chronic condition, where, otherwise, we might have a condition that would gradually produce self-remission. They then decide, well, what we're going to do is, we're going to give everybody a lump sum and tell them they can't come back. And that's what the British do. That's their solution.

Now fast-forward to the 1970s and the debate after Vietnam. Remember, most of the American soldiers are out by 1973, and the war is completely over by 1975. There's a great concern with the psychiatric casualties of the war, and again these debates are renewed, but with a difference. The British said, we can't decide, we can't determine who was traumatized and who's got an authentic memory and so forth, so we're just going to take that for granted and then decide what do to next. The Americans did something quite different. They said, we know what to do next. We have a name for what's wrong with men who have been traumatized and have a disability: it's called a "service-connected disorder." We have a variety of disorders that fall into this category: physical injuries, permanent injuries, temporary injuries, etc. This would be a service-connected disorder, and we have to treat it the way we treat all service-connected disorders, and that is, we will have to conduct periodic examinations, assessments, and ratings. We'll have to create PTSD rating boards, disability rating boards to examine individuals. That's not problematic because that's going to be consistent with what we have been doing.

What is problematic is the question of who's entitled to compensation, and what we've got to do is to develop a standard that will determine who is entitled and who is not. That standard is constructed around the notion of an authentic traumatic memory. What is adopted at that time is a clinical profile with a distinctive inner logic. This inner logic begins with the idea that there is a traumatic event, a terrible event of some sort, an extremely disturbing event that then creates a memory. This memory recurs and recurs, and it's extremely painful and distressful to individuals, so that's the second criterion. The third criterion is that the individual consciously or unconsciously strives to avoid those situations that will stimulate the recurrence of the memory — it's called "avoidance behaviour" — and/or strives consciously or unconsciously to numb himself or herself — most of these are cases of himself — against the emotional effects of memories when they do return. One way in which people numb themselves is a kind of psychological distancing of oneself from one's wife, from children, and others, but an even more effective way, as we know, is with alcohol or with drugs. So substance abuse is redefined as a symptom of PTSD, as an adaptation to the traumatic memory — that's part of the third criterion. The fourth criterion is that there is a physiological arousal that is associated with the disorder, which is partly to be explained by the unconscious anticipation of the recurrence of the traumatic memory, so that the body becomes mobilized, the autonomic nervous system becomes mobilized for fight or flight. You have it all wrapped up as one package.

I describe these criteria as joined by an inner logic because everything that follows is predicated on the relationship between that precipitating event and the installation of the memory. It's the event that creates this indelible memory. Well, at first, it sounds as if it's inevitable. How else could it happen? What alternative would there be? It's a good question, and if you want an answer, what you've got to do is to cross the threshold back to the time before 1980. Go back, for example, to the World War I psychiatrists, who had very good explanations of how the syndrome can, in fact, begin with the distress, with the anxiety, with the depression that the individual feels, perhaps with the substance abuse. Then, following the onset of the syndrome, a cause of the syndrome is sought. This cause is the traumatic event which the sufferer discovers in collaboration with a therapist. This event, at the time it occurred, may not have been a terribly distressful event, but then it gets reinterpreted, in a sense, as the cause of the syndrome.

Now, amongst the many people who wrote about this was Freud. Freud, of course, is now anathema in much of psychiatry, but Freud wrote about this, and he used the German term, *Nachträglichkeit*, to describe the reinterpretation of a memory, so that the memory then assumes a power that the original event did not have. That's another way it can happen. The long and the short of it is that these explanations are all predicated on the idea that memory is remembering, that our memories of the past are not a library of photographs. They are not a collection of videotapes. Again and again in the PTSD literature, that's precisely how traumatic memories are described: as photos, as videotapes — I don't know, maybe today people talk about DVDs — of those memories, as something that is stored away.

DC: The idea of memory as a storage system was the lynchpin of the diagnostic scheme that was introduced into the DSM, the handbook of officially accepted mental disorders, in 1980. Post-traumatic stress disorder would recognize the suffering of many Vietnam veterans and provide a yardstick for compensation. The disorder was defined, as Allan Young has just said, by four criteria: a traumatic event, resulting in a painfully recurring memory, which the sufferer tries to avoid and which produces a variety of distressing and disabling symptoms. These criteria are bound together by what Young calls an "inner logic" imparted by the traumatic memory. Otherwise, he says, many of the symptoms of PTSD could be symptoms of any number of other disorders.

AY: The only way in which you can differentiate PTSD from these other disorders — major depressive disorder, generalized anxiety disorder, substance-abuse disorder, which often go together in the absence of traumatic memory — the only thing that distinguishes them is the fact that there is this inner logic. They're all connected. Some of the symptoms of these other disorders are what, in psychiatry, are called "non-specific symptoms." Some of them are not even, by themselves, psychiatric symptoms: difficulty sleeping, difficulty concentrating, irritability. I'm not criticizing this, but, by themselves, they're not necessarily symptoms. What makes them symptoms of PTSD is that they are all tinctured, they're all coloured by that memory. What connects them together is the memory and this process that I described that connects the event to the memory to the adaptation to the memory and the physiological arousal caused by the memory. That's what glues them all together.

DC: The symptoms of post-traumatic stress disorder become symptoms only when they are glued together by a traumatic memory, but in the years since 1980, this inner logic has been increasingly overlooked. Clinicians will now diagnose partial PTSD, where only some of the criteria of the disorder are met. This has led to a dramatic expansion in the scope of an illness that was at first closely tailored to the circumstances of the Vietnam War. A second, equally important factor has been the ongoing redefinition of what constitutes a traumatic event.

AY: It starts off in 1980 as being very precisely defined, and when I tell you what the definition is, maybe you'll have a picture in your mind of some poor guy slogging in a jungle somewhere in Vietnam, with rocket-propelled grenades going off above his head, and so on. The definition is: an event outside the range of usual human experience — that sounds like it, right? — that will be profoundly distressful to almost anyone, anywhere, anytime. That's pretty unequivocal. It was defined to be unequivocal precisely so that it would be a definition that could not be rejected by the Veterans Administration with regard to awarding compensation and things like that. It makes it quite clear that these are non-trivial events.

What's happened from 1980 on, although there are people within the field who deny it, is that that definition has been gradually broadened. There's been a shift away from an event that is defined objectively by characteristics that are external to the patient. "An event outside the range of normal human experience" doesn't tell you anything about the patient, about the individual. "Profoundly distressful to almost anyone" means to almost anyone. It doesn't tell you anything about the patient. There's been a switch, and the switch is to redefine the traumatic event as an event in which the individual subjectively perceives the event as threatening in some way. A common example, with a lot of claimants, is automobile collisions. You are rear-ended in your car, and you say, geez, I was rear-ended. I saw my life flash before my eyes when this happened. It was such a shock, a surprise, a fright. Well, that could conceivably be, and there would be no problem diagnosing it as PTSD, a trauma-level experience.

DC: The number of people diagnosed with post-traumatic stress disorder has expanded steadily since 1980. A current bulletin of the Canadian Mental Health Association states that one in ten Canadians is affected by PTSD or a related anxiety disorder. The National Institute of Mental Health in the US estimates that 7.7 million Americans will suffer from PTSD in a given year. One of the ways in which the category has enlarged is that it can now pertain to events in which the victim was not personally involved.

AY: We all know what happened on 9/11, and we know that 9/11 was covered intensively on television, and everybody has seen the photographs of the airplanes crashing into the World Trade Center, pictures of people falling. People jumping down from the World Trade Center, the toxic cloud going through lower Manhattan — those images were broadcast throughout the United States and throughout the world. There are PTSD researchers, epidemiologists, who have attempted to study the traumatic effects of those images on populations both within the New York metropolitan area and in other parts of the United States. The results of this research have been published over the last five years in the leading psychiatric and medical journals: the *New England Journal of Medicine*, the *Journal of the American Medical Association*, the *Archives of General Psychiatry*, the *American Journal of Psychiatry*. Those findings are extremely interesting. The people who were interviewed, people who live in Seattle or Broken Mesa, Arizona, and all around the world, were given a list of symptoms. They say, for example, since 9/11, I'm definitely having problems sleeping, as I remember. This is now six months later. I'm more irritable, or something. So they get one symptom, sometimes they get two symptoms, using the model of partial PTSD. From the point of view of this inner logic I've been talking about, this doesn't make any sense. These are non-specific symptoms. We don't know what they mean. We don't even know if they're symptoms. However, those symptoms are then collated, they are brought together, they're aggregated into tables. On those tables, we have the four core symptoms, and the percentages of people who have them. What is being constructed on the page is a very convincing case of PTSD, with all four symptoms, but the individual who is being represented, the person who is being represented, is what I've called a "fictive person." It's not someone who exists. It is a compound, something that's been brought together, aggregated from, in some cases, thousands of individuals to create or construct a kind of a golem.

DC: In PTSD research of this type, symptoms are separated from the individuals who reported them and then added together to create composite sufferers whose only reality is statistical. This represents the culmination of a process that Young thinks has been going on in PTSD treatment and PTSD research for some time. A diagnosis that initially derived from the concrete historical and biographical circumstances of the men who fought in Vietnam has finally become a universal category capable of being recognized and applied without reference to history or biography.

AY: One of the goals of the treatment program, and, I think, one of the goals of the research as well, is to universalize what was a historical situation. What was particular with regard to these men has been generalized to produce, in the end, a language of suffering that is not historically particular. It represents, by 2007, a kind of psychiatric Esperanto that enables us to talk about suffering anywhere. It enables us to take the suffering and the misery of some Vietnam veterans, who trace their psychiatric condition to atrocities that they passively observed or participated in, and to somehow be able to speak about them in the same breath as the suffering of Holocaust survivors — a comparison I believe to be obscene. One gets a kind of clinical equivalence. This is all suffering that can be described in the same clinical terms. In a certain sense, it's the job of the psychiatric language to lift the experiences out of a historical context, a moral context, and to transform the questions of morality into essentially psychological questions, psychiatric questions.

DC: Many of the PTSD sufferers whom Allan Young studied at the Veterans Administration hospital where he did his field research had been involved in events that raised profound questions of justice and morality. Some had participated in the massacre of civilians. Another reported murdering a fellow soldier under cover of battle. And so on. In treatment, these became therapeutic issues. Questions of justice and morality were set aside. "Ideologies of the sort found at this hospital," Young writes in an eloquent passage towards the end of his book, "make truths by eliminating other truths." This brings us back to Young's point that all knowledge is produced within "epistemic cultures," which determine for their members what is worth knowing and how it is to be known. The oracles of the Azande produce witches. Medical science produces post-traumatic stress disorder. Is there then no difference, I asked him, between science and other epistemic cultures?

AY: This is a recurring question within psychiatry: what is distinctive about science? The answer to that is a thoroughly unsatisfactory answer, but it's the answer that we teach here in the Faculty of Medicine at McGill. We call it "the scientific method," and it's the notion of being able to make a hypothesis that is a falsifiable hypothesis and then to attempt to disprove it, and if you cannot disprove it, then you've got reason to be confident in the hypothesis. This is the familiar argument of Karl Popper.

I have another definition. It's a much, much simpler one. That is, I know what non-science looks like, and one of the characteristics of non-science is that it ignores or explains away argument, lack of consensus, and contradiction. One of the characteristics of every epistemic culture — and Ian Hacking has described this beautifully — is that it is, as Hacking says, "self-vindicating," and this is also true of scientific cultures. Self-vindicating means that the people involved can recognize contradictions — they can recognize unexpected results when they are produced in the course of research — but they have ways of explaining those contradictions, and not only explaining them, but appropriating them and saying, well, they're really not contradictions at all. They really represent not a falsification of my original hypothesis, but an elaboration of what my thesis is. They've been very helpful in telling us things that we did not know before. I think this kind of self-vindicating quality within science is inevitable and is certainly necessary. Karl Popper, who first formulated this notion of falsification and falsifiable hypotheses, made, in one of his books, a very interesting observation. He said, if you take literally the doctrine of falsificationism — that you try to falsify a hypothesis, and if you cannot falsify it, no matter how hard you attempt to falsify it, those are grounds for having confidence in it — then what happens if, the first time you try to falsify your hypothesis, you succeed? Am I suggesting that you just get rid of it and look for another hypothesis? And he said, absolutely not. If we look at the most successful of scientific hypotheses, they're constantly falsifying themselves, and it is often the faith of the scientist in the hypothesis, his stubbornness in sticking with the hypothesis, that, in the end, gives us a hypothesis that is useful in a variety of ways. So he said — I'm putting words in his mouth — that the self-vindicating quality is not simply to be disparaged. *But,* and this is my answer to your question, it is also something that needs to be recognized. Scientific facts are, in fact, produced, and one way in which they're produced is within an epistemic culture that has got this self-vindicating quality to it. What is non-science, for me, is not an approach that is stubborn, that holds to a hypothesis that has been falsified, but rather one that ignores the role this process of self-vindication plays in the retention of a falsified hypothesis. There should be at least a marginal skepticism with regard to the power of epistemic cultures to vindicate themselves and to confirm themselves. It's not necessarily a good thing. In the case of Azande witchcraft, it does a great job, and Evans-Pritchard shows this. It predicts things. Those things really happen, they recognize contradictions, they're able to explain the contradictions. Still, in the end, they believe in witches. The argument that I'm making is that the difference between the Azande and a real scientist is a kind of professional skepticism, an awareness that what we're dealing with are epistemic cultures. We don't have some kind of big carpet, and underneath that carpet is reality, and we're gradually unrolling the carpet, and we're discovering a reality, and this reality is a reality that exists before we have attempted to encounter it. Rather, it is a product of that encounter.

DC: The hallmark of science is self-awareness, a recognition that all objects of knowledge are provisional by the very fact that they are produced. Science makes knowledge. It doesn't just, in his image, roll up the carpet and find it innocently lying there. Scientific cultures, like all epistemic cultures, are invariably self-vindicating. Their very coherence depends on their being so, and one cannot eliminate this attribute, only allow for it. And here we come finally to Allan Young's reproach to the world of the PTSD research and treatment: it has become a dogmatic community which resents and rejects criticism. On this point, he knows whereof he speaks, since he has himself sometimes been vilified for his skepticism.

AY: When people do research, let's say, on PTSD, and send the resulting articles in to a journal to be reviewed, and the reviewers of that article are selected from a very, very homogeneous community in terms of what their perceptions are, in terms of what they want to safeguard and what they want to preserve in conventional knowledge, that is unscientific. To the extent that they marginalize critics, to the extent that they stigmatize critics, that is unscientific. What I'm coming to, and my concluding remark, is that, in many respects, these two problems are endemic to mainstream PTSD research.

The Trouble with Physics

LEE SMOLIN

The dominant position of string theory in theoretical physics is a puzzling and disappointing turn of events because there's not a consequence for experiment. Nobody has been able to extract anything that is what we call falsifiable; that is, if it's not seen, then the theory is wrong. This has just never happened before in the history of physics — a remarkable thing. As radical as general relativity was, even before Einstein had it in a final form, he knew what the key experiments were. He knew what he would have to match. It was the same thing with special relativity, the same thing with quantum mechanics, the same thing in every successful instance in the history of physics. The experimental check comes right away. There's always the contact with nature. This idea that some thousand very gifted, very highly placed people in the most elite places in the world passionately believe in something and have worked on it passionately for two decades, without a hint of how to test it experimentally, that's unlike anything that's ever happened before in the history of physics.

— Lee Smolin

During the last twenty-five years, theoretical physics has been transformed by what's sometimes called the string revolution. Once you
get past the basic metaphor of tiny vibrating strings as the ultimate
constituents of matter, the theory is difficult to paraphrase because it's
actually a whole family of theories involving dimensions that can only
be imagined mathematically. But one can say, simply enough, that its
research program now tends to monopolize attention and resources
within theoretical physics. This is a situation which Lee Smolin, himself a theoretical physicist, regards as unhealthy. He expressed his
dissatisfaction in his 2006 book *The Trouble with Physics: The Rise of
String Theory, the Fall of Science and What Comes Next*. There he argues,
first, that string theory is untestable, and, second, that its glamour
disadvantages other equally worthy research programs. However, the
book is much more than just a complaint about string theory hogging
the limelight in theoretical physics. It also takes a wide-ranging look
at the unresolved questions that have perplexed physics for the last
century. And it makes a plea for a return to the more philosophically
adventurous style that Smolin thinks characterized the physicists of the
early twentieth century.

Lee Smolin is a member of the faculty of the Perimeter Institute at
the University of Waterloo. I interviewed him at his home in Toronto
and asked him how his career in theoretical physics had begun. It
started, he replied, with a teenage encounter with the writings of
Albert Einstein. It was the early 1970s. Inspired by Buckminster Fuller,
Smolin had dropped out of school in Cincinnati and was trying to start
a company that would make geodesic domes. He imagined elongating
the basic structure in order to adapt it to various domestic uses — pool
covers, say, or garages — which required him to learn tensor calculus,
the mathematics of curved surfaces. That was how he discovered
Einstein.

LEE SMOLIN

I went to the public library and got some books on tensor calculus, and I
started to study them, and every book had a chapter on general relativity
because that was the mathematics that Einstein used to describe curved
surfaces. I studied general relativity a little, and I found it interesting,

so I went back to the library, and I got a book of essays about Einstein which had one essay by Einstein, his "Autobiographical Notes." It was the only thing he ever wrote that was autobiographical. I was at a personal crisis, a sort of high-school-angst type of personal crisis. My girlfriend had split up with me, the band I was in had fired me, and I was supposed to have a date with another girl, but she stood me up. It was a warm spring evening, so I sat down, and I read that essay. Einstein described why he went into science. He had this vision of the world and human life as difficult and, sometimes, crummy, and — this is a rough quote — by the existence of a stomach, human beings are doomed to chase money and chase material things. At the same time, there's this great, beautiful, transcendental world out there, which is reality, which is space and time and matter, and they have this beautiful mathematics and truth that describe them, and you can transcend your little human life by somehow grasping some of this transcendental reality. I was just hooked. It was exactly right for me at seventeen. I just decided that evening that I would do that, and somehow I had the idea that I *could* do that, and I have no idea why, because I'd never taken a physics course. I had been refused entry into the physics courses at my high school because the physics teacher was a very right-wing guy, and I was this left-wing student activist, and he had told me that he would refuse to let me into any of his courses. I'd never taken physics at all, and I didn't know much about it, except I'd studied some of Einstein's theory of general relativity, but somehow I got the idea that I could do it. So it was very simple. In really just a few hours, I had a kind of mission. I have no idea why I guessed right, but I've stuck true to that ever since.

DAVID CAYLEY

One of the things that inspired Lee Smolin in Einstein's "Autobiographical Notes" was the essay's philosophical breadth. This was a quality he eventually came to admire in the whole generation that created the new physics of the early twentieth century.

LS: My understanding of the history of twentieth-century physics is that there was a revolutionary period from the beginning of the century till the late 1920s when physics was dominated by great revolutionaries: Einstein, Bohr, Heisenberg, Schrödinger, and so forth. These people had a certain style of doing physics which was very oriented towards finding new foundations for physics. It was very much rooted in experiment, but also equally rooted in philosophy and the philosophical tradition. Einstein and these others knew the philosophical tradition very well. They were unembarrassed and unashamed to talk about Kant or Leibniz or Mach in their discussions with each other about physics. They were also very independent; they disagreed with each other. For instance, Einstein and Bohr famously disagreed over quantum mechanics, but they were friends. They enjoyed each other's company. They deeply respected each other. There was also something very European, founded in the style of the European academy, about them.

DC: What these men debated about was the meaning of the discoveries they had made. The new science of quantum mechanics, for example, had created a number of puzzles: light, which had been assumed to be either a particle or a wave, now appeared to be both at the same time; separated particles were shown as still linked with one another in ways that appeared to violate the principle that nothing can travel faster than the speed of light; and things at the quantum level seemed to assume a definite and determinate shape only when observed. At first, the philosophical implications of these findings were urgently discussed, but then these debates were set aside.

LS: The style of physics changed strongly between the 1930s and the 1950s, and there were two reasons for that. One of them was that the centre of physics moved, because of World War II, from Europe to the United States, and so a much more pragmatic American spirit arose and dominated physics in much the same way — and I think there's an interesting analogy to be drawn out here — that the Abstract Expressionists in New York in the fifties stole the dominance of the future of art from Paris and Berlin and so forth. That was one reason. Another reason was that quantum mechanics was in complete form. The older generation were fighting over whether they accepted it. There were deep philosophical issues. Einstein, Bohr, Heisenberg, Schrödinger

— the generation that made quantum mechanics — were continuing to debate and argue over its meaning, its philosophical implications. The younger generation were more pragmatic. They accepted it as given. They were tired of philosophy. They saw the older generation as wasting their time with all these endless philosophical arguments that got nowhere. They just wanted to take the theory and use it, and indeed it was possible to use it to extend the domain of science to the properties of materials, atoms, molecules, nuclei, elementary particles, stars, and galaxies.

There was a huge range of things to be done with quantum physics, and so this new generation was, for both reasons, much more pragmatic. After World War II, funding from governments became more important, and science grew exponentially. The universities grew in this period. Science became much more a profession, much more an industry, and the older, philosophical, foundational, individualistic style didn't fit as well. Indeed, when I was in graduate school and in the period after graduate school, I was educated as a particle physicist by the people who had been triumphant in this new style. They had made, before I went to graduate school, the Standard Model of particle physics, which is the reigning triumph of the pragmatic style. I was very fortunate to have as my teachers the people who made that model, like Steve Weinberg, Shelly Glashow, Sidney Coleman, and others. But the romantic sense that I had about what a scientist was, which came from Einstein and my readings of Einstein and his fellow revolutionaries from that period, didn't really have a place in the world of science that I was introduced to in graduate school. Now, there were some remnants of that world at some smaller universities scattered throughout the world. They were called then the "relativity community." They were remnants, people who were students or associates of Einstein who had been able to establish a few footholds, a few research groups here and there. Eventually, I got absorbed into that world because I fit better into it, but it was definitely a marginal world compared to the dominant style of science.

DC: What Lee Smolin and the "relativity community" wanted to keep alive were the issues that had preoccupied Einstein. Some physicists, like Niels Bohr, had simply accepted the findings of quantum mechanics on the basis that our knowledge, in principle, can never be complete, but Einstein could not accept this limitation on knowledge. He believed that physics should be able to produce a complete picture of the physical foundations of our world, and consequently he ended up opposing what he himself had begun.

LS: See, the paradox of Einstein is that he started everything in the twentieth century. He is the dominant intellectual figure. Quantum mechanics goes back to him. He was the first person who understood the need for quantum mechanics. The duality between waves and particles goes back to him. Photons go back to him, as well as special relativity, as well as general relativity. So twentieth-century physics is the consequence of his work, and, at the same time, he dissented from its form when it finally coalesced into the current form of quantum mechanics in the late 1920s. He dissented from it because he believed strongly that quantum mechanics, giving only statistical predictions, could not be a real theory. A real theory must tell you what's happening in every individual process, not just give you statistical predictions. He had other reasons as well. Quantum mechanics doesn't give a picture of reality in the absence of our intervention, which he felt science was about. Science was about going to this transcendent viewpoint and seeing the world as if we were absent, and quantum mechanics does not do that. Quantum mechanics is a kind of code for describing the effect that we have on the world when we make controlled experiments. The mathematical pieces of quantum mechanics refer to our interventions, our measurements, and that was completely unacceptable for Einstein.

On the other hand, for Bohr, who came out of a philosophical tradition through Kierkegaard, Schopenhauer, and so forth, this was how science had to be. For Bohr, science was an extension of our interaction with the world and our conversation with each other about the world. Thus they had a profound disagreement.

Finally, physics, after this pragmatic turn, when the revolution was prematurely declared over, was left, so to speak, on hold. There is quantum physics, which takes you from the most fundamental particles through solids and the properties of materials and which takes account of three of the four forces we know about. The fourth one, which it doesn't touch, is gravity, and gravity, according to Einstein's theory of general relativity — and we have good evidence for this — is an aspect of the geometry of space and time. So there's a deep issue about the nature of space and time which is left unresolved and can only be resolved by finding one theory which describes nature as a whole, which incorporates space and time and gravity, the cosmos on the large scale and quantum physics: atoms, particles, and so on. This is what we sometimes call the problem of "quantum gravity," and this was unresolved and, for a long time, was not very much worked on, except for a few individuals scattered from here to there. Indeed, that was the problem that I took on from Einstein, that I then began to work on as an undergraduate and in graduate school, and that I have continued to work on.

DC: The problem of quantum gravity on which Lee Smolin has worked arises from the incompatibility between Einstein's theory of relativity and quantum mechanics. Relativity describes the geometry of space-time, of which gravity is an aspect. Quantum mechanics describes the fine structure of matter. But the two don't fit together. The key issue has to do with the nature of space and time.

LS: The discussion really goes back to the time of Newton, and it goes back to debates between Newton and his followers and Leibniz and his followers. Now, Leibniz was a philosopher and a mathematician. He was a rival to Newton for the honour of having invented the calculus. He was also a diplomat. He attempted to reconcile the Protestants and the Catholics and reunite Europe, and he functioned as a diplomat in the Hanovers' going and taking the English throne. You find him all over the place. There are some philosophers who call him the smartest person who ever lived. He anticipated symbolic logic. He anticipated computers. And he had a deep disagreement with Newton about the nature of space and time, which is easy to explain. For Newton, before there's any matter, there's space. Space exists absolutely, and particles move within it. It's like a bare stage in a empty theatre. The stage is there when the play is not on. Then the audience comes in, the actors come on, and they play the play. The stage directions tell them "stage left," "stage right," "back," and so forth, but the stage is always there, no matter what's going on, and positions on the stage are absolute and unrelated to the play. For Newton, space was the sensorium of God. Space was God and was God's way of knowing what was in the world — by feeling things through their positions in space. Newton saw space that way, as absolute, and time in the same way. There was some absolute time. It's meaningful to say what time it is in the universe.

Now, Leibniz was also very theological — he had this big project of reconciling the religions, the Christian religions, with each other — but he had a view of religion that led to a view of space and time in which there was no absolute framework. Space was only a matter of relationships between things, and without things to have relationships, there would be no space. His picture is more like street theatre. Until the actors come and start playing, there is no stage, there is no space. The actors create the space that they play in. That, we say, is a relational view of space. If I talk about where I am, I can give you an address, which describes where I am relative to the centre of Toronto, to a grid of streets and so forth, so this is all describing position relatively — in terms of relationships. Newton believed that, behind all of that, there's something else, there's some absolute notion of where one is. Leibniz denied that, and that was the basis of their disagreement.

Scientifically, Newton's view won out at the time. It was much easier to establish a workable physical theory based on Newton's conception than on Leibniz's conception. But the debate was kept up from that period into the end of the nineteenth century by philosophers and occasionally by physicists, and Einstein was a student of this literature. What Einstein did in general relativity was to fully realize the Leibnizian view: general relativity is an entirely relational view of space and time. There is nothing to space and time but a dynamically evolving network of relationships. However, the Newtonian view remains in the rest of physics, which means in quantum physics. In all the other treatments of physics, space and time are treated essentially as Newton treated them, as an absolute background. The old language for this debate is the absolute versus the relational view of space and time. The contemporary vocabulary we use, say, in arguments over the future of physics, is between background-dependent theories, where you specify what is the background of space and time, and you fix that background, and then you have things moving against that background, whether they be particles or strings or fields; and, on the other hand, background-independent approaches, where there is no fixed background, and space and time, if they exist at all, are part of a dynamic network of relationships. This continues to be the centre of debate. For example, this is the key issue in the debate between string theorists and essentially everyone else who thinks about the problem of quantum gravity.

DC: Background dependence versus background independence is the key underlying issue in contemporary theoretical physics. String theory has retained the Newtonian framework. For all its fantastic machinery of extra dimensions, Smolin argues, it still assumes a stable framework within which its strings move. He has opted for Einstein's view of things and attempted, along with other colleagues who have adopted this minority position, to integrate quantum mechanics into a relational account of space and time. This has produced a number of new theories, each different, but with a certain family resemblance among them all. The variant Smolin works on is called "loop quantum gravity," and it pictures space-time as an infinitesimally small gridwork or graph.

LS: The picture of relational space that arose — it's very Leibnizian — is imagined as a network, just like the Internet or the phone network. There are nodes and there are connections between them, and that makes a kind of graph that you could draw as a picture hanging in space. That's a picture of the quantum geometry. When you look closely at atoms, they have a discrete structure. There are the orbitals and so forth. The discrete structure of the quantum geometry of space consists of these graphs, and that's actually a mathematical theorem. I certainly was not able to prove it, but better people proved it. We've understood how to talk about the geometry of space and time and their dynamics completely in quantum-mechanical language, and it works.

DC: One of the crucial elements of this theory is that it gives space a discrete structure, "discrete" here meaning made of separate and discontinuous bits. This is what it adds to the theory of relativity, which pictures space-time as smooth and continuous.

LS: When you apply quantum mechanics to matter, then you discover that atoms have discrete structure. When an atom absorbs some energy, say, with a photon coming from the outside, the photon is a discrete thing which is absorbed, and the atom jumps to another energy state, which is a discrete state. So what quantum mechanics implies in many cases is that there's a discreteness in nature.

Now, in general relativity, the geometry of space is something continuous and dynamical. If you think of a surface, maybe the surface of a body of water, then the geometry of space is like a three-dimensional version of the surface of water. There can be waves through it, it oscillates, and so forth. When we apply quantum mechanics to it, we discover that there's a discrete structure. If you take a closed container — say, this coffee cup with a lid on it — and you ask how much volume is there, in the classical world, you would get a number, which would be a continuous number. In the loop quantum gravity world, if it's correct, then there's a counting that you actually do. There actually are little pieces of volume. There's a smallest possible volume of a piece of space. String them together, and you get bigger volumes. In the picture of graphs, a node of the graph is one of these little elemental volumes. We can compute the units involved, and they're very, very small. There are about 10^{20} — so twenty orders of magnitude, one with twenty zeroes — of these elementary lengths inside the nucleus of an atom, so the scale is tiny. Nonetheless, the theory predicts the magnitude of these elementary pieces of volume and elementary pieces of area. A particle moving smoothly through space, if you looked down at it on this scale, would look something like a chess piece making discrete moves on a chessboard.

By the way, Einstein anticipated this idea. It's not very well known — John Stachel, the historian, talks about "the other Einstein" — but Einstein, in some letters and writings, did anticipate the structure of space turning out to be discrete in this way, although he didn't know how to work on it himself. He even says, "I don't know the right mathematics to get started on this." He was so frustrated by his search for a unified field theory, in which he was using continuous mathematics and continuous geometry, that he anticipated this.

DC: The theory of loop quantum gravity can be distinguished from the model which currently dominates theoretical physics, string theory, in a number of ways. One important difference is that the strings of string theory are imagined as moving in a continuous space, a space that lacks the discrete structure that Lee Smolin has been talking about. Another is that loop quantum gravity is a testable theory, whereas string theory is not. If he and his colleagues are right, then the structure they attribute to space should produce minute variations in the speed at which light travels — an effect physicists call "dispersion." Once the predictions of the theory are refined, this effect should be observable.

LS: This is a tiny effect, and it used to be said that you couldn't measure any such effects because you would need an elementary particle accelerator the size of a galaxy or something like that. But, it turns out, we have access to experimental apparatuses much huger than a galaxy: we see photons that travel across the observable universe from ten billion years ago, when they were created in big explosions called "gamma-ray bursts," and these gamma rays are detected, roughly one a day, by satellites in earth orbit; that creates a laboratory the size of the universe. If our expectation is borne out at the right order of magnitude, then there should be a difference of about a thousandth of a second in the time it has taken these photons to spread out. That's a small time, but for the modern electronics that go into these satellites, that's trivial. Computers operate many, many orders of magnitude faster than that, so this becomes a measurable effect. A satellite called GLAST which is studying these gamma-ray bursts was launched in June of 2008, and over a few years, it may discover such an effect. There's a handful of such effects, which are amplified by long, long travel times across the universe, and which amplify these tiny differences that we couldn't detect in an experiment on earth. These, if you like, are very small modifications of Einstein's principles of relativity that come from the discrete nature of space and time that loop quantum gravity predicts.

DC: The theory of loop quantum gravity that Lee Smolin has been laying out here is just one of a number of ongoing efforts to solve a fundamental problem of contemporary theoretical physics: the incompatibility between quantum mechanics and the general theory of relativity. String theory is another such attempt, but, in Smolin's view, it has come to exert an unhealthy sway over the entire field. This has happened since 1984, when crucial problems in string theory were solved, and it took off, developing, in the process, what he calls an almost "cult-like atmosphere," as well as a dominant position in many of the most important research institutions. This worries him, first, because string theory makes no testable predictions and, second, because he thinks it has developed into a form of groupthink that is inhibiting the free and independent spirit that should properly characterize theoretical physics. He makes this argument in his recent book, *The Trouble with Physics*, and he begins his development of the argument here with an explanation of what string theory is.

LS: String theory is a hypothesis about the unification of all the forces and all the particles in nature. As it's described in string theory, there is this string moving through fixed space or, it turns out, sometimes it can be a higher dimensional object moving through a higher dimensional space. If you imagine it has different overtones of vibration, like a guitar string or a violin string, then you can arrange it so that the vibrations correspond to different kinds of elementary particles. But then there's a trick, and the trick is, if you want to get the elementary particles to come out looking like they do in our world, then you have to accept the existence of these extra dimensions. So string theory comes as a package deal: if you buy it, you have to buy that there are six extra dimensions, but you can fold them up into any of an infinity of different geometries. The strings wiggle at different frequencies, depending on the geometry of these extra dimensions in which they run around. By adjusting the geometry of the extra dimensions, you can get any kind of particles you like, practically speaking. You can make them more massive, or less massive, and allow for different interactions. The triumph of string theory — it's a beautiful idea — is that you can capture all of elementary particle physics in the vibrations of these strings. The Achilles heel is that the only way you have a hope of getting elementary particle physics to come out right is by exploiting the freedom to choose the geometry of these extra dimensions. The problem is that this is an infinite freedom.

DC: Let me try and paraphrase to see if I've understood you. You have your observed phenomena. To explain them, you make up a story, but there's no way of testing whether it's the right story. You might as well say that it's microscopic elves that are generating reality.

LS: Well, there's more structure in string theory than in the theory of elves, but it *is* a puzzling and disappointing turn of events because, in the original excitement back in 1984, there was a sense one was closing in on something unique. At that time, it appeared that, if you kept the geometry of the ten dimensions of space and time completely flat and unadorned, then there were only five possible versions of the theory, and there was a belief that maybe these were the only five ways a world can be. But if you take this question of the geometry of the background seriously, then you see how the fact that the theory is defined against a background is its Achilles heel; the fact that you can choose the background arbitrarily to be any of an infinite number of backgrounds means that the theory is actually an infinite number of different theories. As far as we know, there's not a consequence for experiment. Nobody's been able to extract anything that is what we call falsifiable; that is, if it's not seen, then the theory is wrong. This has just never happened before in the history of physics — a remarkable thing. As radical as general relativity was, even before Einstein had it in final form, he knew what the key experiments were, he knew the data, he knew what he would have to match. There are things that Newtonian physics didn't get right, and he had several predictions that could be tested with the technology of the time within a few years — two years, three years. It was the same thing with special relativity, the same thing with quantum mechanics, and the same thing in every successful instance in the history of physics. The experimental check comes right away. There's always the contact with nature. It's very easy to make mathematical structures, to invent mathematical games that make no contact with reality. A situation in which some thousand very gifted, very highly placed people in the most elite places in the world passionately believe in something and have worked on it passionately for two decades, without a hint of how to test it experimentally, that's unlike anything that's ever happened before in the history of physics.

DC: Smolin regarded string theory, when it first appeared, as a promising conjecture, and some of his own ideas were formed in his attempt to understand it. But somehow this conjecture has turned into an academic industry which monopolizes resources and marginalizes alternative approaches. In his book, he offers several explanations of why this has happened. One is that string theory's weakness — its elasticity, its being an infinite number of theories — is, at the same time, its strength. Making no testable prediction, it can never be disproved. Another is that the pragmatic turn in physics has produced problem solvers when what was needed was philosophers. Finally, he argues, the hegemony of string theory has been fostered by the way science is now organized and funded.

LS: Because there are few funding bodies and many universities, there's been a uniformitization of judgment. A university or university department can't make judgments on its own any more. They rely, more and more seriously and more seriously than in the past, on the judgment of panels and funding bodies and the judgments of what are called "visiting committees," who are eminent scientists from the most prestigious universities who come and visit them and tell them what fields they should be working in. There's much more of a steering of directions by consensus, which is generated nationally and internationally. It's more dominant in the larger countries, like the United States, and less so in the smaller countries, like, say, Canada or Holland, just because they are smaller. That's the first thing.

  Another thing that happened is, the universities expanded expo-nentially between the 1940s and the 1970s, and in this period, a lot of government funding from the various federal governments came in and drove the growth of science especially. During that period, it wasn't difficult to get an academic job, so there was a lot of decentralization. You got a PhD. I know people from that period, most of whom are now retired, who never applied for a job. They never had any career anxiety. If you wanted to be a professor at McGill or Harvard or some other elite institution, there was still competition, but it wasn't hard to find a place for yourself somewhere. There was a kind of decentralization, and because of that decentralization, there was, as there should be in the academic world, a variety of viewpoints, of approaches to any problem.

Then, all of a sudden, the expansion stopped in 1972 — I think the date is pretty precisely known — and there was this cadre of young PhDs that had been growing and growing and growing, and they all just, like, went over a cliff. All of a sudden, there were no academic positions. This certainly hit science very hard. I think it hit all fields to some extent. All of a sudden, there was a very quick transition to an atmosphere of scarcity in which many more PhDs were being trained than could fit. In that atmosphere of scarcity, there came to be a lot of professionalization and competition among the universities to keep their funding and to acquire funding by hiring faculty who would do the best job of bringing funding to their universities. As with anything, I think, there is a variety of ways of being a good scientist. However, now that there was a lot of competition and professionalization, a narrower idea developed of what constitutes a good scientist. Now you had to make difficult judgments. You had to compare people. The funding bodies had to reject most of the applications and so forth. The model that got established is that a good scientist is a very clever problem solver, and to be a very clever problem solver implies that you're working in what Thomas Kuhn called the paradigm, in which there's a well-defined theoretical basis, a well-defined experimental practice, and a well-defined set of techniques, and you're incrementally improving the techniques within a well-defined domain. That's what Kuhn called "normal science."

A term for this kind of scientist that I like comes from Eric Weinstein, who's a mathematician and economist. He says that such people are "hill-climbers." Imagine the problems in science as hills. You're trying to find a theory, and a good theory is a hill of a certain height. The right theory is a mountain somewhere over there. Everything is shrouded in fog. There are people who are good hill-climbers; you put them down on the side of a hill, and they'll go up. Basically, if you define who you want to hire as whoever climbs the hill the fastest, given that the hill is well defined, then that's what you'll look for. However, then you end up with communities of people stuck on the tops of hills, yelling at each other because all they know is the technology involved in climbing their hill, insulting each other from hilltop to hilltop, when the real thing is some mountain shrouded in fog way over there. To get to the mountain, you need people who are what Eric Weinstein calls "valley-crossers," these crazy people who go down when everybody else is competing to go up the fastest, because they don't like crowds, or they have their own idea, or they smell something which maybe is a valley or a forest over there, and they go off on their own. Of course, many of them get lost, but they're people with an incredible talent for somehow finding new mountains. You put them down on the side of a hill, and they'll go down, but they'll end up higher than everybody else somewhere else a few years later. Watson and Crick were such people. They wandered into molecular biology. They were physicists. They had no training. They didn't know anything. They had never worked with the techniques involved. They just started playing, putting these things together, and they did it; otherwise, they would have been total failures. Einstein was such a person. That's why he had trouble getting an academic job. Bohr was such a person. He was two-thirds a philosopher who had strange ways of talking.

My hypothesis is that not enough thought has been given to making sure that there's space for the valley-crossers or for the rebels. And the thing that's poignant is that there aren't that many such people. Supporting all of them would not be that expensive because in fundamental physics, there are maybe fifty people like this, and that might even be an over-estimate. In biology, there are maybe a few hundred people like this around the world who have trouble getting their laboratories started. If I think it's bad in physics, then it's worse in biology, where the success rate for an assistant professor of biology trying to get a first grant from the NIH in the United States is twelve percent. That used to be much higher. The successful ones tend to go in with a big, already established research program, and it's no wonder that departments who need their faculty to have funding to make their balance sheets work go for the people who are following the big, established directions. Now, I don't have the knowledge to say that, therefore, progress in medical science is much slower than it would be, but what I think I believe is that progress in fundamental physics is much slower than it should be.

Scientific ideas, scientific research programs are like new firms, and one would like the funding bodies to think more like venture capitalists. These are investments. If you talk to venture capitalists, they understand that, if you want to keep the front of technology moving, most of the things that you invest in will fail. For every Microsoft, for every successful biotech start-up, there are twenty that fail. But there's no choice. That's what you have to do. If you just kept putting money into Microsoft, the progress of technology would stop. You have to keep funding the start-ups, and you have to accept the high risk that goes with a high rate of progress and a high rate of return. String theory is a high-risk, high-payoff thing, and, in some sense, the message of that part of my book is, that's great, but so is quantum causal history, so is causal sets, so is twister theory, so are Gerard 't Hooft's ideas, so is Alain Connes's noncommunative geometry. What you need is a portfolio of these things.

DC: All these things you have named are . . .

LS: . . . are approaches to quantum gravity . . .

DC: . . . which have a much lower profile.

LS: Yes.

DC: And much less support . . .

LS: Yes.

DC: . . . than string theory.

LS: Yes.

DC: And that's why most people listening to us will never have heard of them.

LS: Yes, and most of them involve half a dozen, a dozen, twenty people around the world, and that's about right. One of the things that happens, when you start to get thousands of people working on something, is that almost all of them are redundant when it's a new, speculative idea. Many young people complain about that. Many of the young string theorists are highly gifted people who believe in what they're doing, who know what they're doing and know why they're doing it. But in quiet moments and in the corners, so to speak, one still hears complaints from them that their research program is too directed, that there's too much faddishness. If there's some problem which is unsolved for three years, then the support to work on it gets dropped. People who continue working on it don't get faculty positions, don't get invited to the key spots at conferences which identify who the young stars are, and, therefore, the problems don't get solved. Even within string theory, my advice is to have a much more venture-capital point of view: support things. Don't support anything too much.

DC: According to Smolin, the predominance of string theory in the strongholds of theoretical physics is a product of two great forces: the pragmatic turn in American physics, which marginalized philosophical concern, and the current organization of the academy, which fosters conformism and monoculture. His prescription calls for a revival of the intellectual independence and philosophical breadth that he thinks characterized early twentieth-century physics. Physicists, until well into the nineteenth century, were still called "natural philosophers," and Smolin, in a sense, is arguing that that remains their proper vocation.

He also thinks that physics and physicists need to engage with other elements of culture. Natural science, for him, is part of culture and ought to be exchanging ideas with the arts, with the social sciences, and with the humanities. His books, *The Life of the Cosmos*, *Three Roads to Quantum Gravity*, and *The Trouble with Physics*, which we've been discussing, are attempts to situate science at what he calls the "front of contemporary culture." But the participation of other scientists in this conversation remains blocked by a narrowness that he thinks is still all too prevalent.

LS: There are many scientists in my field — and I assume it must be the same in other fields — who know very little outside that field and know nothing of the history or literature of their field going back to before the beginning of their careers. That is unfortunate. I think that you can have some people like that, because we need a variety of people, but having too many people like that impedes progress because they get stuck in ruts, they get stuck at the top of hills and they can't get off.

Still, the conversation that I'm trying to foster is going on, and what those of us in it are seeking is somehow to extend this conversation across the whole front of culture. It's not my place to say it, because I'm involved in this, but I have the impression that science is part of this front of culture. As culture evolves and culture progresses, those of us who are at the front — scientists, artists, social theorists, architects, writers, and so forth — have a lot to say to each other, and I think that maybe we lose out with the over-specialization and departmentalization. And there are venues for that conversation. Some of them are conferences. Some of them are friendships. Some of them are cities. The city is a venue for conversation. That's what they're for. As a New Yorker who moved to Toronto, I'm pretty excited about Toronto and the people I meet here — writers, people in theatre, films, and music, people in technology, people in politics. Toronto is more like New York than it is like London and Paris. It's a more open, accessible city. You could be in Paris forever and never meet anybody who does anything different than you, whereas in New York, once you're somehow in New York, you're continually meeting people who do something interesting other than what you do. I'm finding that Toronto is like that as well. I think I'm in a very lucky position because I've been fortunate to be able to write books as part of this community, and, through the writing of the books, the circle of people that I'm fortunate to have as friends, that I communicate with, has grown much, much wider. It has been well worth it, certainly worth every hour of time and agony that went into the writing, which, I'm sure you appreciate, is considerable.

From Knowledge
to Wisdom

NICHOLAS MAXWELL

There are these two absolutely basic problems: to learn about the universe and ourselves as a part of the universe, and to learn how to create a civilized world. Essentially, we've solved the first problem. We solved it when we created modern science. That is not to say that we know everything there is to be known, but we created a method for improving our knowledge about the world. But we haven't solved the second problem. And to solve the first problem without solving the second problem is very, very dangerous. The crisis is this: the crisis of having science without civilization.

— Nicholas Maxwell

Science has been very successful at producing knowledge, but knowledge without wisdom, or science without civilization, is a dangerous thing, according to Nicholas Maxwell. He believes that the reason we have the one without the other is that science, as now practised, does not question its own purposes or investigate its own presuppositions. It transforms the world but cannot transform itself. Therefore, he says, "we need a revolution in the aims and methods of academic inquiry, so that the basic aim becomes to promote wisdom by rational means, instead of just to acquire knowledge."

Nicholas Maxwell is a philosopher of science, now retired from University College, London, and the author of *From Knowledge to Wisdom*, first published in 1984 and reissued in a revised edition in 2007. I came across him by happy chance when one of his titles caught my eye during a browse through the philosophy shelves in the University of Toronto's science library. Even among the several acres of books arrayed there, Maxwell's *Is Science Neurotic?* jumped out at me as a title that posed an interesting and novel question. So I took it down and discovered that, yes, indeed, according to him, science is neurotic. A neurosis, as Freud defined it, is the tension produced by some repressed aim. In the famous Oedipus complex, the boy child can't face up to his animosity towards his father, whom he sees as a rival for his mother's love, so he represses his problematic desire to get rid of his father and puts a more acceptable face on his purposes. Science, Maxwell says, suffers from what he calls rationalistic neurosis. It pretends to be a disinterested pursuit of knowledge, which neither makes nor grants any assumptions. In actuality, however, it must take many things for granted. These range from basic requirements — like the money that supports the enterprise and the strings that are sometimes attached — to more rarefied suppositions — for example, that nature actually possesses the simple, unified, law-like structure that science has ascribed to it. Science, in other words, pretends to be more reasonable than it is, and in the process, it prevents itself from actually becoming more reasonable.

I was intrigued by Maxwell's diagnosis, and after reading more of his work, I arranged to interview him at his home in London. I began by asking about the origin of his interest in science.

NICHOLAS MAXWELL

Well, I suppose I'd have to go back to my childhood, to when I was four. I know I was four because of the house where we were living. One evening I was lying in bed, and I started to think about where space ends. After a bit, I thought, well, yes, of course, if you go far enough, eventually you must come to a vast wall, and that's the end of everything.

Then, of course, the awful thought occurred: what is behind the wall? Okay, I hadn't discovered a theory or anything, but I had discovered a serious scientific problem — the fundamental problem of cosmology. Then, I remember, when I was about six, I was walking in the garden where we were living in Surrey, and I was thinking about the problem of why things are the size that they are. I had the thought that this must be because the ultimate constituents of matter, of material, must have a certain definite size. There must, in other words, be atoms to fix the size of other things, like the size that we are, or cats are, or trees are, because otherwise things wouldn't know what size to be. I would never have been able to articulate it or put it into words — it was almost as if I felt it rather than thought it — but I remember thinking at the time that this was very profound, and I remembered it. Then, I remember sitting in the living room one evening — I suppose this might be more philosophy — and suddenly realizing that my gaze wasn't, as it were, going out and being in touch with the room around me. All that was happening was that light was coming into my eyes and causing me to have the experience of seeing, so that in some sense this living room that I was seeing must be inside my head. But how on earth could the living room be inside my head? I remember being absolutely and utterly baffled by this.

DAVID CAYLEY
These early experiences convinced Maxwell of his scientific vocation. His aim would be to understand the very nature of things.

NM: Up to adolescence, my passionate desire was to be a theoretical physicist. I wanted to do what Einstein had wanted to do. I wanted to find out what kind of universe this is. I felt, when I was around nine and ten, that to live and die and not know what kind of universe you're in is the most unspeakable tragedy. It means your entire life is a complete waste. I thought of history in those terms — just a record of disaster, with all these people who had lived without knowing even the most basic things about what kind of world they were in. I just thought, this mustn't happen to me. I must find out. I remember telling myself, whatever happens, you must never betray this all-important quest.

DC: This quest developed a new dimension with adolescence. The mysteries of time, space, and matter yielded to the mysteries of human personality.

NM: Then I decided that what really mattered was life and what was happening in the hearts and souls of people, and the way to get at that was through literature. I was lucky that quite a lot of the world's literature was on the shelves at home, in translation. I was able to read Tolstoy and Dostoevsky and Kafka and Virginia Woolf and Stendhal, which I did, and I felt this was all about my world. I could recognize all this. Now my great aim was to be a novelist.

DC: Maxwell never did become a novelist or a physicist, though both impulses fed his approach to the discipline on which he finally did settle: philosophy. He went to the University of Manchester to study it.

NM: After my first year at Manchester, the great explosion in my life happened. I came to the conclusion that this desire of mine either to be a theoretical physicist and discover the secret of the universe, or to be a novelist and discover the secret of life, when pushed to the limits of absurdity, was really the desire to be God. And that this was really not just a bit impractical, but also undesirable. It was much more important that I was myself, but I didn't know what it was to be myself. It had never occurred to me to think that I was of any significance whatsoever. I also remember — this seemed to me to be the key thing — writing down the sentence, "The riddle of the universe is the riddle of our desires." This came out of all this stuff about the absurdity of wanting to be God. I concluded that instead of seeking to know the universe and, in a certain sense, to master it through knowledge, the real task should be to discover what is of value in life. The real problem is: what do we want? What, ultimately, is of value?

Then I went back to Manchester and I thought, well, I must tell all my philosophy lecturers and professors about this enormously important discovery that I've made — because I'd also concluded that philosophy should be about how we're going to live, not about knowledge. I discovered I couldn't really even open my mouth. The rest of my working life has been a long effort to discover how to open my mouth and speak.

DC: The desire for knowledge, Maxwell saw, is, by itself, just appetite, unless it is governed by wisdom — the ability to discern what is of ultimate importance. This realization would eventually become the cornerstone of his philosophy of science, but it would be some time before he would fully understand where the insights gained during what he calls his "great explosion" were leading him. A crucial step on the way was his encounter with one of the twentieth century's most influential philosophers of science, Karl Popper.

NM: I learned something of enormous importance from Popper. The one thing that really struck me was something he says in *Conjectures and Refutations*, one of his books. He's arguing for the importance of realism, and he says that without realism — without the view that the world exists independently of us, that there are the facts of the case — one can't really have something like justice. The idea of justice can survive bad judges and bad courts and bad decisions, but what it can't survive is the idea there is there is no such thing as the facts of the case, that, in the end, this person really is either guilty or not guilty. If you don't believe in objective facts, then there is only going to be political pressure and whose story wins out. The whole idea of justice disappears. It was really a moral argument that convinced me of the importance of realism.

Up till then, in an attempt to understand what had happened during this explosive summer that I'd experienced in 1961, I'd felt that I'd got this great message for the world: that we should be our own prophets. So here was I, a great prophet, saying that we should all be our own prophets. This seemed to me contradictory — a sort of nonsensical message. I came to the conclusion that, in the end, there were only different stories, and that one shouldn't take the story that science tells us about the nature of the world as the true story when it was just one story alongside other stories. And it was really reading this little passage of Popper's that convinced me that this was no good and that one had to see the world as something that's existing there, independently of us, something we can improve our knowledge and understanding of.

DC: Karl Popper's defence of realism steered Maxwell away from a view that would later become quite influential in the history and philosophy of science: that science is, after all, just a story. He was important to Maxwell in other ways as well, so, before we went on, I asked him for a précis of Popper's account of science.

NM: His fundamental idea is that in science one can never verify theories. You can only falsify theories. But falsification, though it sounds a very negative thing, is actually what makes scientific progress possible. In science, when a theory is falsified experimentally, we can discover that we're wrong. We know that it can't be right. We have to find something better, or attempt to find something better. That's how it is that science makes progress. All our knowledge, in the end, is conjectural knowledge. Popper generalized this to rationality as a whole, arguing that reason has to do with being critical. Falsification is an especially severe form of criticism. Observations and experiments in science are part of the activity of criticizing ideas, criticizing theories. They're a very severe form of criticism, because if the experiment is decisively at odds with the predictions of the theory, then however beautiful the theory may be, the theory has to go by the board — at least that's the idea.

Criticism, without the backup of observation and experiment, may be less decisive, but it's also enormously important from the point of view of improving our ideas. There's been this long-standing problem in philosophy about skepticism: how do you refute skeptical arguments? In a certain sense, Popper transforms that argument. Far from skepticism being the enemy of science and reason, it's the essence of the scientific attitude and of rationality. To be skeptical is to be rational. To be rational is to be skeptical, and it's through being skeptical that we learn. We learn through making mistakes and trying to improve on our mistakes. Popper applies this idea of critical rationalism, which is generalized from science, to social and political problems, especially in a famous book called *The Open Society and Its Enemies*. The idea, really, in that book, is that a rational society is a society that encourages and tolerates criticism. But for that you need a plurality of ideas and ways of life and values, so that a rational society is an open society, a society that can sustain and tolerate different ways of life. That, for Popper, is what civilization is or what it should be.

DC: Maxwell first came across Karl Popper as a reader, but later he would get to know him personally. Popper had fled his native Vienna in 1937, just ahead of the Nazi invasion of Austria, and after the war he had settled in Britain, where he taught at the London School of Economics. Maxwell joined his seminar there in the 1960s.

NM: Popper had been described to me as this funny man who would say, "Ve must not be dogmatic." And he would say it very dogmatically. "Ve must refute ourselves." You know, how absurd. He was regarded, and he was described to me by my professors at Manchester, as a bit of a buffoon. But after I read this passage in *Conjectures and Refutations*, I started to read whatever else I could get hold of, and for a while I became an occasional student at the LSE where Popper was presiding. I went along to his seminars and was immensely impressed. This was the first person I had encountered in academia who I thought was speaking . . . every time he opened his mouth, it seems to me, he said something extraordinarily interesting and at odds with what everyone else was saying. I learned an enormous amount. There was a passion and a fierceness about him — and he could be quite fierce.

I was a convinced Popperian, and I breathed a great sigh of relief, because I felt that, in a certain sense, what I'd been trying to do had been done already. Popper had done it, and I could relax. But there were a few things that he hadn't sorted out, and one thing in particular I have tended to see as my fundamental problem. It's the problem which goes back to my childhood and to my two early vocations, the scientist and then the novelist. How do we put the scientific vision of the world together with the world as we experience it, the world of literature and art and values? That I saw as an absolutely fundamental problem, and it was really that problem that haunted me.

DC: In trying to put together the world of science and the world of experience, Maxwell was taking up a problem that Popper had not really addressed, but he also, in time, developed differences with Popper over some of the questions Popper had taken up. Despite being, as he says, a convinced Popperian, he came to think that Popper was quite wrong in believing that a scientific theory stands or falls by its ability to explain the observed facts.

NM: I think where Popper fundamentally went wrong is that he took the view — the quite standard view — that the basic aim of science is truth, and the basic method is to assess claims to truth — theories — empirically and impartially with respect to their empirical success and failure. In some ways, Popper is highly unorthodox, but in this one respect he's highly orthodox. He accepts this very basic, standard empiricist picture of science: that the basic aim is truth, the basic method is assessment with respect to the evidence. It seemed to me that this view just didn't correspond to what actually goes on. And it couldn't. It couldn't actually work. The problem arises all the time in physics, and in a very exposed, naked form. Physicists are all the time choosing theories that are simple or unified or explanatory. The fact of the matter is that whatever theory one considers — Newtonian theory, Einstein's theory of general relativity, quantum theory, or any other theory of physics — all these theories run into empirical difficulties somewhere. By arbitrarily putting things together into a sort of patchwork quilt theory, one can always concoct ad hoc rivals that would more successfully fit the experimental results and would indeed make new predictions. This new, horrible theory — horribly lacking in simplicity and unity and explanatoriness — would actually be much more empirically successful. But these horrible theories never get considered at all, even though they're much more empirically successful than the theory that we accept. To me, what that means is that the whole enterprise of physics is making a big assumption about the world. It's assuming, a priori, that all these horrible theories are false because if some of them were true, then we would have to look at them. I'm not just saying that this is what actually goes on, but that something like this has to go on because there are always going to be endlessly many theories which will equally well fit the available evidence. We have to have some extra considerations in addition to empirical considerations to select out the one or two theories that we want to take seriously.

The whole scientific enterprise has to make a big assumption about the nature of the universe, which, in the end, amounts to saying that the universe is more or less physically comprehensible. It must assume that there is some kind of underlying unified pattern, a more or less unified pattern, and that there are physical laws governing all phenomena. Therefore, the theories that we need to consider are the theories that are more or less unified, or that contribute to the overall unified picture that we have, as well as being empirically successful. So there are these two requirements. What has happened, I think, is that science has misidentified its basic intellectual aim. It isn't just truth. It's rather to improve our knowledge of the physical comprehensibility of the universe — the universe being presupposed to be physically comprehensible, or more or less physically comprehensible. There is this highly problematic aim built into the methods of physics, built into the whole enterprise of physics. It's there in practice and it has to be there.

DC: Scientific theories, in Maxwell's account, are selected according to certain assumptions, assumptions which are unprovable and therefore, as he says, problematic — they might be wrong. They are, in a sense, articles of faith, without which science wouldn't be possible at all. But what he calls the standard empiricist picture of science denies that there are any such assumptions. A theory is true if it accounts for the evidence, and that's that. This empiricist claim, because it claims too much, can sometimes provoke critics to deny the objectivity of science altogether. Maxwell decided that what was needed was a philosophy of science that could systematically spell out the entire hierarchy of assumptions on which the practice of science rests.

NM: I came to the conclusion that once one has recognized that physics, and therefore, in a sense, the whole of natural science, has this deeply problematic aim, this deeply problematic assumption built into the aim — that the universe is more or less physically comprehensible — then the problem is to try to provide a framework for improving our ideas as we proceed. The best way to do this is to represent our aims in the form of a hierarchy of aims, or a hierarchy of assumptions. As one goes up the hierarchy, the assumptions get less and less substantial, and so more and more likely to be true, more and more required to be true for science or for knowledge to be possible at all. These higher-order assumptions are also less and less likely to need revision. Then, as you go down the hierarchy, the assumptions and the associated aims get more and more specific, more and more substantial, and therefore more and more likely to be wrong, and therefore more and more likely to need revision. You create a framework of reasonably stable assumptions and associated methods, and, within this framework, you can revise your much more specific and much more uncertain assumptions and associated methods as you proceed in the light of success and failure. I think this is how science has actually proceeded.

Nevertheless, I thought at first that no scientist would ever believe this account because it's too much at odds with the official view. It seems to imply that there is an article of faith in science, and scientists always want to distance themselves from religion and say there's no article of faith. Everything is just evidence. But then it dawned on me that in a way Einstein had believed in something like this, because he'd once said, for example, that the most incomprehensible thing about the universe is that it is comprehensible. So I started to read Einstein, and I came to the conclusion that I'd been anticipated, and that he'd more or less held a view like this one.

DC: Arranging the assumptions underlying the practice of science in a hierarchy allows one to distinguish between what changes and what endures. The search for a unified pattern of physical laws has characterized science for centuries. New discoveries might put today's Standard Model of particle physics out of date by tomorrow. One of the advantages of a scheme that clearly marks this difference, Maxwell argues, is that it highlights the overall continuity of science.

NM: The whole idea of the hierarchy is to try and keep the revisions as low down in that hierarchy as possible. Our hope is that we've got it right if we go far enough up the hierarchy. We're probably right when we say that the universe is such that we can continue, for a while, to acquire some knowledge of our local circumstances. Let's hope that's right. And let's hope that the universe is such that we can improve our methods of improving knowledge. And let's hope that the universe is in some way or other comprehensible, maybe even physically comprehensible. But almost certainly it isn't physically comprehensible in the more specific way that we currently think it is. Indeed, it's an implication of the idea that the universe is physically comprehensible in some unified way that our current fundamental theories must be false. Physical comprehensibility implies that the true theory is going to be a unified theory. In principle — not in practice, but in principle — it will predict all physical phenomena. But our current theories are very dramatically not unified. On the one hand, there is Einstein's theory of general relativity and his theory of gravitation, and on the other, we have what physicists call the Standard Model, which is the quantum field theory of particles and their fundamental forces. These two theories are horribly not unified. They're very different. They don't fit together. Furthermore, the Standard Model itself is, in a way, three different theories stuck rather artificially together. This is why so many theoretical physicists these days work on string theory, because they think that it can unify these two theories. It is really the drive for unification that's behind all this research on string theory, behind the hypothesis that there is only one kind of string, it vibrates in different ways, and that's what produces the apparently different particles.

DC: String theory answers the need for a unified explanation of the phenomena of physics. It is pursued out of a deep conviction that there must be such an explanation, but, according to Maxwell, this must finally be a matter of faith rather than knowledge. It's what he calls a problematic aim, and science generally has not wanted to face up to the fact that it rests on such metaphysical foundations. The consequence is that science has become neurotic. It is, as we say today, in denial.

NM: For scientists, I think, the big problem is that to acknowledge this does violence to the official view about the nature of science, especially when one comes to defend science against other things, like religion and politics. What is it that is so special about science? In science, nothing is taken as an article of faith. Nothing is taken on trust. Everything is open to being assessed empirically, and that's an extremely simple line to take. In politics, there is dogma of various kinds. In religion, there is a book, there is an oracle of some kind or other. There is faith. In science, there isn't. But if I'm right, and if you acknowledge that, for science to be possible at all, you have to make these highly problematic assumptions, then there is in a sense an article of faith, and it's an article of faith that you can't do without. That simple way of distinguishing science from religion and science from politics doesn't work anymore.

Of course, there's another way of doing it, but then you have to say that the difference is that in science these basic assumptions are subjected to constant scrutiny and that we are always seeking to modify and improve the assumptions we make in the direction of that which seems to be the most fruitful from the point of view of helping us to improve our empirical knowledge. That's really what marks out science from these other things. That doesn't go on at present because of the neurosis of science, but that's what one should be able to say about science.

DC: Science is properly understood as the continuous practice of self-criticism, the potentially endless tuning and refinement of its own assumptions. Yet science, broadly speaking, understands itself as an unproblematic pursuit of knowledge. It denies that scientific theories conform to rational as well as to empirical requirements, claiming that science is built entirely from the ground up. And this mistaken philosophy, in Maxwell's view, has had eminently practical consequences because it has been, in effect, built into the institutions of science.

NM: I think it's very important, when we talk about the philosophy of science, to think not just of the ideas of philosophers of science, or the ideas of those scientists who talk about philosophy of science. One needs to look at the philosophy of science that is, as you say, built into the institutional enterprise of science. What matters is textbooks and the training that scientists get. This operates in all sorts of ways — for example, in criteria for publication in a scientific journal. I have found it extremely difficult to get my ideas across to the scientific community because they're not testable. The very thing I'm trying to criticize — namely, the idea that only testable ideas should be considered in science — excludes what I'm trying to say. The very thing I'm trying to criticize protects itself from criticism. It's like a lobster pot. Once you go in, it's very difficult to get out again. I knew slightly a famous scientist who was quite sympathetic to ideas like mine, but he felt he couldn't say anything about this in his teaching because he would ruin his students' chances of getting jobs. Their careers would be damaged. Scientists may personally feel that they don't believe in the current orthodox view, but they nevertheless have to take it into account when they're trying to get a paper published — they have to write it up in such a way that it will be accepted. After all, you know, all scientists need to get publications — not just for the proper reasons of communicating their ideas, but to advance their careers, get research money, and so on. All these scientific activities, all these aspects of the scientific enterprise, are informed by a philosophy of science. And then, of course, there is this discrepancy between what really goes on and what is supposed to be going on.

DC: This discrepancy, in Maxwell's opinion, tends to breed theories which deny the rationality of science altogether. Science plainly isn't what it says it is — so its productions must actually reflect underlying social and political realities. An example of this way of thinking is the "strong program" in the sociology of scientific knowledge. First put forward by members of the Science Studies units at the Universities of Bath and Edinburgh in the early 1970s, it claimed that all scientific knowledge can be given a social explanation. It's a position that Maxwell deplores.

NM: I think the strong program, and what is sometimes called a social constructivist approach — the idea that science doesn't really provide us with knowledge, it provides us with a kind of myth, and we change our myths every now and again — has been absolutely disastrous for the history and philosophy of science and for understanding science. It's driven a wedge between the philosophy of science on the one hand and the history and sociology of science on the other because philosophers of science on the whole don't accept this view. They think it's nonsense, whereas many historians and sociologists of science do accept it. I think it's all fundamentally misconceived as well. I mean, there is a sense in which it stems from the feeling that the fundamental problem in the philosophy of science — how do we make sense of the scientific enterprise? How do we see it as a rational endeavour? — is unsolvable. It is simply not the case that evidence determines choice of theory, so therefore you have to bring in other things, like social factors, to explain why certain ideas are taken up rather than other ideas. It is not rational considerations that govern the choice of theory, but these various social factors. Well, I think actually that my work provides a solution to this problem.

DC: Science is governed by rational considerations, but it won't admit that it is, and this has left the field open for sociological explanations. Maxwell does not, in fact, deny that science is pushed and pulled by all sorts of social and political imperatives. He just wants to establish that it answers to rational criteria as well. That admitted, he is entirely willing to recognize that science has a social structure, and that denying that structure is also part of the scientific neurosis.

NM: In what we've been talking about so far, we've been really talking about metaphysics and the buried metaphysical assumptions of science — the untestable assumptions. There are also assumptions that have to do with values — for instance, the assumption that what we want is not just explanatory truth but important truth, valuable truth. There's also a political dimension to science, because science gets used to do things, and this is even more problematic. I mean, values are problematic enough — what do we mean by important? important to whom? when? where? and so on — but there are even more problems when it comes to the aim of using science. The ideal, of course, is that it will be used to promote human welfare, but, as we know, science has actually been used in all sorts of other ways. As an integral part of science, we need to be looking critically not just at the metaphysical assumptions of science, but also at the value assumptions that influence the directions in which scientific research is being pursued. We also need to be looking at the use to which science is put. But it goes further than that, and here I'm walking in a direction parallel to Karl Popper's, as we discussed earlier. Popper expanded his falsificationist conception of scientific method to become a general idea of critical rationalism. Similarly, I want to expand my hierarchical conception of scientific method to become a general idea of rationality, according to which, in life quite generally, whenever our aims are problematic, we need to represent them in the form of a hierarchy, so that we can improve our aims and methods as we proceed, as we live, in the light of our successes and failures. This idea, it seems to me, applies above all to the goal of trying to create a better world. This goal is profoundly problematic, and it's not just how we get there that's the problem, but where we should be trying to get to in the first place. What do we mean by a better world, a good world? Most of the ideas that have been put forward in the past have either been absolutely appalling or unrealizable or some combination of the two. When you come to the attempts to achieve them — from extreme kinds of socialism to national socialism — the twentieth century is littered with the horrors of utopian endeavours. In my view, this is because we have failed to build into our political and other social institutions something comparable to my conception of scientific method, a method by which we can learn to recognize the inherently problematic character of our basic aims and by which we can build the institutional means to improve our aims and methods as we proceed.

DC: Maxwell readily concedes the nightmarish character of most previous attempts to render society more scientific, but he thinks this has resulted from a fundamental misunderstanding. The Enlightenment and its many sequels did preach the power of science to improve society, but science was generally taken to mean the accumulation and application of knowledge. What should have been emphasized, he says, were science's methods, not its accumulated products. And even here, he is speaking more of a spirit of critical rationality than of a precise set of techniques.

NM: The situation in social life is very, very different from what it is in science. In science you can do experiments, and as long as you don't blow people up or anything too dreadful, no one is hurt if the theory fails. If the theory doesn't work, it dies, and it doesn't matter. But if you try doing experiments in the social world, then whatever it is that you've created acquires a life of its own. If you decide we need a certain kind of institution and you decide to create it, even if it's just a political initiative of some kind or a new policy, it tends to acquire a life of its own. The people who are now operating the institution won't be happy if you say, a year or two down the line, look, this has been a complete fiasco, we must dismantle the whole thing, because they'll suddenly be out of a job and out of a career. They will fight for its continuation and give distorted explanations as to why this is a good thing to be doing. So it's extremely difficult to conduct experiments. I mean, there are all sorts of problems associated with trying things out in the social world that don't tend to arise when you're trying things out in science. This is why it seems to me to be so important that we try, as far as we can, to develop the capacity to try out these experiments in our imaginations and critically assess them in our imaginations. In other words, we need to develop very accurate and highly critical imaginative experiments. We should surround our actions with many imagined actions, considering what might have happened if we'd done this, or if we'd done that. Of course, there is a certain sense in which experiments are being done all the time because the world is very big and very diverse, and people are doing all sorts of different things. My picture of academia is that academia ought to be our attempt to, as it were, hoover up and develop and critically assess our very best ideas about how to solve our problems of living at all levels — at global levels, at local levels, at individual levels. It should act rather as civil servants are supposed to act for governments, only, instead of acting secretly, academia should be acting openly for the public. This means that it's really important that it not be necessary to have a PhD in order to get a good idea taken up and developed and scrutinized. It shouldn't be necessary to be an academic. Somebody in some village in India may have found a solution to a problem that people face in other places, and it may be working extremely well. One of the tasks of academia should be to seize upon such things, to seek them out and then broadcast them and make them available, but also to scrutinize them and to see whether they really are working.

DC: Maxwell's proposal for the reorientation of academia is summed up in the title of his book *From Knowledge to Wisdom*. It is no longer enough to use the critical methods of science to improve our means when what is obviously deficient is our ends. The question about knowledge should not be just, is it true?, but what is it for? This is the revolution in science that he is calling for.

NM: If you take the view that the basic aim of science, the basic intellectual aim, is truth and the basic method is to assess claims to knowledge impartially with respect to the evidence, then it isn't really a part of the intellectual aspect of science to question its aims in any larger sense. That doesn't really come into it. Whereas if you take the aim to be valuable truth, then it immediately does. It's absolutely apparent. Is this of value? Is this the kind of knowledge that we should be seeking to acquire? It stares you in the face. You cannot possibly teach an undergraduate course or a course in science in school without raising these questions. In fact, these questions do get raised, and this is why I say there is a discrepancy between the official story and what actually goes on. People are more intelligent in practice than they are in principle. But officially, there aren't the mechanisms, and scientists who cry out about the abuses of science are liable to be somewhat ostracized or thought to be somewhat unreliable, rather than regarded as just doing the straightforward, plain business of what any scientist should do.

DC: Science must become self-critical, able to scrutinize its own aims as carefully as it now sifts evidence or assesses the truth of theories. It is this self-critical science that Maxwell thinks can also be applied to what he calls problems of living. However, by this he emphatically does not mean the reorganization of society according to some positive program generated by science. This was exactly the problem with earlier projects for the scientific reform of society — from scientific child-rearing to scientific socialism — and why his own proposal might be misunderstood.

NM: I think most people's image would be of a kind of nightmare. If the idea was to extract scientific rationality from science and then impose it on the social world, then we'd all end up as automata, obeying reason in everything that we do. Reason would be a great dictator, and we'd just be obeying. There's a wonderful novel by a Russian named Zamyatin which in translation is called *We*. It's a satire on the Soviet Union, but it's also a satire on this whole idea of the rational society. No one has a name. They all just have a number. They all live deeply rational and totally enslaved lives. In their lunch breaks, they have to link their arms together five at a time and sing songs to Reason as they parade down the street. The story is all about a man who gradually rebels and thinks this isn't quite the way to live. It's a vision of nightmare, you could say. In my view, one of the immense achievements of Karl Popper has been, in a certain sense, to dispel that nightmare. He thinks of reason, not in terms of prescribing what you should do in every single detail, but rather as being critical. Well, that's a important thing to do, even if there are some circumstances in which criticism, and some ways of being critical, can be counterproductive. That's a limitation on Popper's notion of rationality, but the point is that it doesn't tell you what to do. It tells you that you might doubt what you're intending to do and that there may be alternative possibilities. The whole idea is that criticism is only possible if there is a multiplicity of ways of life and values and ideals and ideas. Popper's rational society is the open society, it's not a society governed by reason in that sense. All I'm doing is improving on that. I'm not dispelling that aspect of it.

DC: Rationality of this kind leads to pluralism, not to a tyranny of reason. Science, for Maxwell, is pre-eminently the capacity for doubt, disagreement, and self-questioning, but this is not the account science has so far given of itself. In consequence, our world is poised on the brink of destruction — because we have knowledge without wisdom.

NM: There are these two absolute basic problems: to learn about the universe and ourselves as a part of the universe, and to learn how to create a civilized world. Essentially, we've solved the first problem. We solved it when we created modern science. That is not to say that we know everything there is to be known, but we created a method for improving our knowledge about the world. But we haven't solved the second problem. I mean, we've made progress, but essentially we've not solved the second problem. To solve the first problem without solving the second problem is very, very dangerous because, as a result of solving the first problem, we have enormously increased our power to act, or at least the power of some of us. Science and technology lead to industrialization, population growth, modern methods of travel, modern methods of agriculture, and it's all these immense successes that have created all these other problems of global warming, the extinction of species, the destruction of tropical rainforests, and the immense discrepancies of wealth and of power. What I think isn't appreciated at all is that the crisis behind all the others is the crisis of having science without civilization, without having even grasped what it would be to have a society that has learned how to solve the second problem of civilization, the problem of gradually making progress towards a more civilized world.

Acknowledgements

My thanks to everyone who was interviewed for "How To Think About Science," the CBC radio series on which *Ideas on the Nature of Science* is based. A request out of the blue for a lengthy interview cannot always have been welcome, but I was graciously received wherever I went, and my questions were answered with remarkable thoroughness and eloquence. The success the series subsequently enjoyed with listeners is a testament to this generous public spirit. My thanks also to Barbara Brown of CBC Merchandising who authorized the production of a transcript of the series and to the irreplaceable Ian Godfrey, who did the transcribing, along with Mike Housego. *Ideas* executive producer Bernie Lucht encouraged me to do the series, supported me as it grew to an unprecedented length, and then vetted a draft of every episode with me. My wife, Jutta Mason, helped me in countless ways through the nine months the series was in production. And, finally, I would like to thank Laurel Boone, who edited the book for Goose Lane and made the work of preparing it for a literary public a pleasure.